D0984707

613.96 KOM
Komisaruk, Barry R.
The science of orgasm /

PALM BEACH COUNTY
LIBRARY SYSTEM
3650 SUMMIT BLVD
WEST PALM BEACH, FLORIDA 33406

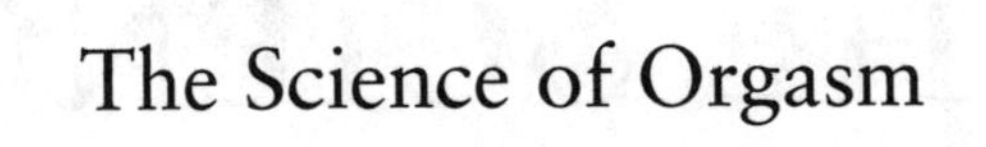
The Science of Orgasm

THE SCIENCE OF

Orgasm

Barry R. Komisaruk, Ph.D.

Carlos Beyer-Flores, Ph.D.

Beverly Whipple, Ph.D., R.N.

THE JOHNS HOPKINS UNIVERSITY PRESS
Baltimore

© 2006 The Johns Hopkins University Press
All rights reserved. Published 2006
Printed in the United States of America on acid-free paper
9 8 7 6 5 4 3 2 1

The Johns Hopkins University Press
2715 North Charles Street
Baltimore, Maryland 21218-4363
www.press.jhu.edu

Library of Congress Cataloging-in-Publication Data

Komisaruk, Barry R.
The science of orgasm / Barry R. Komisaruk, Carlos Beyer-Flores, Beverly Whipple.
p. cm.
Includes bibliographical references and index.
ISBN 0-8018-8490-X (hardcover : alk. paper)
1. Orgasm. I. Beyer, Carlos. II. Whipple, Beverly. III. Title.
RA788.K66 2006
613.9′6—dc22

2006013040

A catalog record for this book is available from the British Library.

I think the most unfair thing about life is the way it ends. I mean, life is tough. It takes up a lot of your time. What do you get at the end of it? A death. What's that, a bonus? I think the life cycle is all backwards. You should die first, get it out of the way. Then you live in an old age home. You get kicked out when you're too young, you get a gold watch, you go to work. You work forty years until you're young enough to enjoy your retirement. You go to college, you do drugs, alcohol, you party, you get ready for high school. You go to school, you become a kid, you play, you have no responsibilities, you become a little baby, you go back into the womb, you spend your last nine months floating . . . You finish off as a gleam in somebody's eye!

—Sean Morey

Contents

Preface

We wrote this book because orgasm is a remarkable phenomenon and one of life's most intriguing experiences. We take a different perspective from that of recent books addressing various aspects of sexual behavior and orgasm, such as those by Bancroft (1989), Kothari and Patel (1991), Paget (2001), Rodgers (2001), Bodnar et al. (2002), Margolis (2004), Hyde (2005), and Lloyd (2005), and from that of the review articles by Meston and Frohlich (2000), Meston, Levin, et al. (2004), and Mah and Binik (2001, 2005). We address the body systems that act in concert to produce orgasms; discuss various ideas about the function of orgasms and why orgasms feel good; define the many types of orgasms, such as "nongenital" orgasms; and address sex differences and similarities, effects of aging, pathologies of orgasms, effects of hormones and drugs, and the neurology and brain control of orgasm.

Philip Teitelbaum commented in a conversation with one of the authors that the brain is like a house, and it is the same house regardless of which window you look into—behavioral, neurological, anatomical, or pharmacological. There are such diverse ways of understanding the phenomenon of orgasm—from the poetic to the medical—that each perspective can argue that the others are hopelessly narrow. The medical approach fails to appreciate the emotional effect of the altered states of consciousness and spiritual aspects of orgasm, while those who view orgasm from the latter approach demur that the nuts and bolts approach is hopelessly limited. We recognize that each perspective is valid, and yet each has its limitations.

For their help in developing and preparing this book, we are happy to acknowledge our many colleagues and friends worldwide, and our family members, for their discussions, suggestions, critical comments, inspiration, encouragement, and direct help. We especially thank Colin Beer, Angelica Breton, Vincent Burke, Arlene Feldman, Gabriela Gonzalez-Mariscal, Kimberly F. Johnson, Adam C. Komisaruk, Christine McGinn, Gina Ogden, MariaCruz Rodriguez del Cerro, Jay Rosenblatt, Alexa Selph, Bill Stayton, Linda Strange, Dennis Sugrue, Alexander Tsiaras and the staff of Anatomical Travelogue, Mary Ann Ulrich, and Jim Whipple.

On a more personal note, Beverly thanks her husband, Jim, children, Allen and Susan, and grandchildren, Kayla, Travis, Valerie, William, and Elyse, for their continued love and support. Carlos thanks his wife, Josefina, and daughters, Maria Emilia and Gaby. Barry thanks his sons, Adam and Kevin, for their interest, love, encouragement, and support, and in remembrance, his wife, Carrie.

The Science of Orgasm

1

Definitions of Orgasm

Although many may prefer that orgasms remain indescribable and undefinable, some have proposed a definition:

"[From the Greek meaning] to swell as with moisture, be excited or eager" (*Oxford English Dictionary,* Simpson & Weiner, 2002a)

"The expulsive discharge of neuromuscular tensions at the peak of sexual response" (Kinsey et al., 1953)

"A brief episode of physical release from the vasocongestion and myotonic increment developed in response to sexual stimuli" (Masters & Johnson, 1966)

"The zenith of sexuoerotic experience that men and women characterize subjectively as voluptuous rapture or ecstasy. It occurs simultaneously in the brain/mind

and the pelvic genitalia. Irrespective of its locus of onset, the occurrence of orgasm is contingent upon reciprocal intercommunication between neural networks in the brain, above, and the pelvic genitalia, below, and it does not survive their deconnection by the severance of the spinal cord. However, it is able to survive even extensive trauma at either end." (Money, Wainwright & Hingburger, 1991)

"An explosive cerebrally encoded neuromuscular response at the peak of sexual arousal elicited by psychobiological stimuli, the pleasurable sensations of which are experienced in association with dispensable pelvic physiological concomitants" (Kothari & Patel, 1991)

"A peak intensity of excitation generated by: (a) afferent and re-afferent stimulation from visceral and/or somatic sensory receptors activated exogenously and/or endogenously, and/or (b) higher-order cognitive processes, followed by a release and resolution (decrease) of excitation. By this definition, orgasm is characteristic of, but not restricted to, the genital system." (Komisaruk & Whipple, 1991)

"A variable, transient, peak sensation of intense pleasure, creating an altered state of consciousness, usually with an initiation accompanied by involuntary rhythmic contractions of the pelvic striated circumvaginal musculature, often with concomitant uterine and anal contractions, and myotonia that resolves the sexually induced vasocongestion and myotonia, generally with an induction of well-being and contentment." (Meston, Levin, et al., 2004)

What exactly is an orgasm? Almost everyone would agree that orgasm is an intense, pleasurable response to genital stimulation: penile physical stimulation in males and clitoral or vaginal physical stimulation in females. Although orgasm characteristically results from genital stimulation, there are many reports that other types of sensory stimulation also generate orgasms, stimulation perceived as both "genital" and "nongenital."

For example, there are documented cases of women who claim they can experience orgasms just by thinking—without any physical stimulation. Their bodily reactions of doubling of heart rate, blood pressure, pupil diameter, and pain threshold bear out their claim (Whipple, Ogden & Komisaruk, 1992). Men and women with spinal cord injury have described that the skin near their injury is hypersensitive to touch—painful and intensely aversive to touch if accidentally brushed, but when stimulated in the right way, by the right person, capable of producing pleasurable orgasmic feelings that may be perceived as emanating from the genitals. A woman with a complete spinal cord injury at the upper thoracic level, whose area of hypersensitivity was the neck and shoulder, claimed to have orgasms from stimulation of the skin of her neck. In the laboratory, her heart rate and blood pressure increased markedly during self-application of a vibrator to her neck-shoulder junction, and she described experiencing an orgasm accompanied by a "tingling" sensation in her vagina (Sipski, Komisaruk, et al., 1993).

In reports by Kinsey et al. (1953), Masters and Johnson (1966), and Paget (2001), women stated that they experienced orgasms from breast or nipple stimulation. Paget (2001) also described orgasms produced by stimulation of the mouth or anus in women and men. Women with spinal cord injury described experiencing orgasm from stimulation of the ears, lips, breasts, or nipples (Comarr & Vigue, 1978). Xaviera Hollander (1981), author of *The Happy Hooker,* said she experienced an orgasm when a police officer placed his hand on her shoulder. In the world of fiction, the heroine of the novel *Kinflicks,* on realizing she has just had an orgasm when her lover held her hand, says she can experience orgasms from stimulation anywhere on her body (Alther, 1975).

The occurrence of orgasms stimulated from many different parts of the body is also reported by persons using marijuana. And cocaine users claim that the rush they feel just after injecting the drug feels orgasmic (Seecof & Tennant, 1986).

Is the sensation of orgasm different in women and men? Vance and Wagner (1976) performed a carefully controlled study in which college students taking a course in the psychology of sexual behavior wrote descriptions of their own orgasms, and a group of judges tried to guess which descriptions were written by men and which by women. The judges were female and male obstetrician-gynecologists, psychologists, and medical students. Before submitting the descriptions to the judges, Vance and Wagner substituted gender-neutral words for specific words in the students' written descriptions (e.g., *genitalia* for *penis* or *vagina*) to intentionally conceal the sex of the writers. The study authors built additional clever and appropriate safeguards into the experimental design, a model for this type of research.

Vance and Wagner's statistical analysis showed that no type or gender of judge was better than any other at discerning the sex of the writer. They concluded that "individuals are unable to distinguish the sex of a person from that person's written description of his or her orgasm . . . Furthermore, neither sex was more adept at recognizing characteristics in descriptions of orgasm that would serve as a basis for sex differentiation, if, indeed, there are factors in such descriptions which can be differentiated."

The students' descriptions are vivid, as demonstrated by quotations selected at random from nine of the forty-eight participants in the study. In each instance, as noted above, the judges could not correctly identify whether the writer was male or female (we don't know, either).

> A sudden feeling of lightheadedness followed by an intense feeling of relief and elation. A rush. Intense muscular spasms of the whole body. Sense of euphoria followed by deep peace and relaxation.

> Feels like tension building up until you think it can't build up any more, then release. The orgasm is both the highest point of tension and the release almost at the same time. Also feeling contractions in the genitals. Tingling all over.

> There is a great release of tensions that have built up in the prior stages of sexual activity. This release is extremely pleasurable and exciting. The feeling seems to be centered in the genital region. It is extremely intense and exhilarating. There is a loss of muscular control as the pleasure mounts and you almost cannot go on. You almost don't want to go on. This is followed by the climax and refractory states!

> The period when the orgasm takes place—a loss of a real feeling for the surroundings except for the other person. The movements are spontaneous and intense.

> A heightened feeling of excitement with severe muscular tension especially through the back and legs, rigid straightening of the entire body for about 5 seconds, and a strong and general relaxation and very tired relieved feeling.

> Basically it's an enormous buildup of tension, anxiety, strain followed by a period of total oblivion to sensation then a tremendous expulsion of the buildup with a feeling of wonderfulness and relief.

> The feeling of orgasm in my opinion is a feeling of utmost relief of any type of tension. It is the most fulfilling experience I have ever had of enjoyment. The feeling is exuberant and the *most enjoyable* feeling I have ever experienced [emphasis in original].

> A building of tension, sometimes, and frustration until the climax. A *tightening* inside, palpitating rhythm, explosion, and warmth and peace [emphasis in original].

> A complete relief of all tensions. Very powerful and filled with ecstasy. Contraction of stomach and back muscles.

These characterizations of orgasm emphasize the physical feeling. A more abstract characterization was reported in a study of more than four hundred male and female university students by Mah and Binik (2005): "orgasmic pleasure and satisfaction were more consistently related to the cognitive-affective characteristics of the subjective orgasm experience than were the sensory characteristics. [The findings support the] salience of the overall psychological intensity of the orgasm experience over the perceived anatomical location of orgasmic sensations."

2

Different Nerves, Different Orgasmic Feelings

An understanding of the sensory pathways that are probably activated during orgasm can suggest the bases for the experiences of orgasm. The pelvic nerve provides the afferent (sensory) nerve supply of the vagina, cervix, rectum, and urinary bladder (Komisaruk, Adler & Hutchison, 1972; Peters, Kristal & Komisaruk, 1987; Berkley et al., 1990). Activation of this nerve can generate orgasm when stimulated vaginally, so it is not surprising that when activated nongenitally (e.g., rectally), the pelvic nerve can also generate orgasm in both women and men. In women, stimulating the rectum in addition to the clitoris, vagina, and cervix could add to the quality—complexity, intensity, and consequently pleasurableness—of orgasm.

One man described a ten-year history of feelings resembling sexual orgasm and ejaculation after each defecation or forceful urination, followed by a rise in pulse rate and a sense of relaxation that changed to extreme fatigue (Van der Schoot & Ypma, 2002). Conversely, in women, uterine, cervical, and vaginal stimulation during childbirth has been reported to induce feelings of the urge to defecate. Thus, "cross-talk" or "referred sensation" or "equivalence" can exist between the feelings generated by vaginal and by rectal stimulation, most likely because the same nerve—the pelvic nerve—carries sensory information from both organs.

In men, sensory activity originating from the prostate (via the hypogastric nerve) during ejaculation contributes to the pleasurable sensation of orgasm, as evidenced by the finding that prostatectomy may diminish this feeling (Koeman et al., 1996). The contributory role of this afferent activity to orgasm could help account for the experience of orgasm in men receiving mechanical stimulation of the prostate during anal intercourse, which would add to the sensory activity via the rectal component of the pelvic nerve.

The hypogastric nerve also conveys sensory activity from the uterus and cervix (Bonica, 1967; Peters, Kristal & Komisaruk, 1987; Berkley et al., 1990; Giuliano & Julia-Guilloteau, 2006; Hoyt, 2006). The orgasmic role of afferent activity via the hypogastric nerve known to occur in men can help account for the parallel drawn in women between the feelings generated during childbirth and orgasm (Newton, 1955). Stimulation of the pelvic nerve may also occur with stimulation of the area of the G spot (the area of the female prostate gland) and may also account for the reports of orgasm and "female ejaculation" from the urethra experienced by some women (Perry & Whipple, 1981; Ladas, Whipple & Perry, 1982, 2005).

The orgasm-inducing effect of breast or nipple stimulation may be due to sensory activity from the breast traveling ("projecting") to certain of the same group of neurons in the brain that receives sensory activity from the genitals (for a review, see Komisaruk & Whipple, 2000). Specifically, there is a cluster of neurons that comprise the paraventricular nucleus of the hypothalamus (*nucleus* is the name given to a cluster of neurons in the brain or spinal cord). These neurons produce and secrete oxytocin, a "neurohormone" (so called because it is a hormone secreted by neurons rather than by endocrine glands), into the bloodstream, brain, and spinal cord in response to breast or nipple stimulation as well as in response to vaginal, cervical, or uterine stimulation. The oxytocin released by suckling stimulates the contraction of the myoepithelial cells (*myo-* meaning "muscle") that envelop the milk-secreting glands in the breast. When these cells contract under the influence of oxytocin, they forcibly expel milk, in the "milk-ejection" or "milk-letdown" reflex. The oxytocin can concurrently stimulate the uterine smooth muscle to contract. In a parallel process, during childbirth, uterine contractions that push the fetus against the cervix stimulate sensory fibers in the pelvic nerve that project along the spinal cord to the same paraventricular nucleus of the hypothalamus, stimulating the neurons to release oxytocin into the bloodstream. This is termed the "Ferguson reflex" (J. K. W. Ferguson, 1941). This oxytocin released during childbirth can also produce expulsion of milk from the breast in women who are lactating.

Since the final common pathway for oxytocin secretion is mainly the paraventricular nucleus of the hypothalamus (Cross & Wakerley, 1977), breast, nipple, cervical, and vaginal afferent (sensory) activities evidently converge on this neuronal cell group. There is normally a significant release of oxytocin into the bloodstream within one minute after orgasm in healthy women, and in some cases blood levels are still elevated after five minutes (Cross & Wakerley, 1977; Carmichael, Humbert, et al., 1987; Carmichael, Warburton, et al., 1994). As described below, activation of the paraventricular nucleus of the hypothalamus has been observed during orgasm (Komisaruk, Whipple, Crawford, et al., 2004), so it is likely that the activity of these neurons, the neurons to which they project, and the neurons that project to them is involved in the pleasurable sensation of orgasm (and the description by some women of nursing and childbirth as generating "orgasmic" feelings).

As described by Ladas, Whipple, and Perry (1982, 2005), the sensory quality of orgasms differs in relation to the part of the sexual system that is stimulated. Vaginal stimulation–induced orgasm is described as involving the whole body, whereas clitoris-induced orgasm is more restricted to the region of the clitoris. One woman described her orgasms produced by cervical stimulation as a "shower of stars." For our research we employ a unique stimulator that consists of a stimulating rod attached to a diaphragm that covers the cervix. Women using this device tell us that when they pull outward with the stimulator, thus pulling on the diaphragm surrounding the base of the cervix, the outward motion produces a very pleasurable suction on the cervix that they have never experienced before.

The difference in sensory quality of stimulating the clitoris, vagina, or cervix is most likely due to the different nerves receiving sensory activity from each of these regions. That is, the clitoris is supplied mainly by the pudendal nerve, the vagina mainly by the pelvic nerve, and the cervix mainly by the hypogastric, pelvic, and vagus nerves. While stimulating each of these regions can by itself produce orgasms (of differing feelings), the combined stimulation of all three regions has an additive effect, producing more encompassing orgasms, or what is described as "blended orgasm" (I. Singer, 1973; Ladas et al., 1982, 2005).

In her recent book, Elisabeth Lloyd (2005) raises an interesting challenge: she contends that women's orgasm has no known biological function: "women can and do get pregnant easily without having orgasm during copulation, and there is no evidence that shows that orgasm makes a difference to fertility . . . birth rates . . . or reproductive success." Furthermore, given that women do not always have an orgasm during sexual intercourse, "under the common assumption that the capacity for orgasm is designed as an adaptation to encourage and reward intercourse, this infrequency must be seen as a design flaw." Lloyd contrasts women's orgasm with men's: "male orgasm is known to be necessary for male reproductive success, while the same is not true for women."

Lloyd places women's orgasm in the same category as men's nipples—that is, a by-product of a structure that has a function only in the other sex. In other words, she considers women's orgasm to be a "by-product" (her term) of men's orgasm. "Males get nipples as a developmental or embryological consequence of females needing to have them," she writes. "And the trait persists among males because it is consistently selected in females." By analogy, since the clitoris and the penis develop from the same embryonic tissue, "Females get the [clitoral] erectile and nervous tissue necessary for orgasm in virtue of the strong, ongoing selective pressure on males for the sperm delivery system of male orgasm and ejaculation."

Lloyd also dismisses any adaptive role of the orgasm-induced peak release of oxytocin in women. Oxytocin has been shown to increase uterine peristaltic contractions, which could thereby accelerate the transport of sperm through the uterus to the site of fertilization in the fallopian tubes (Wildt et al., 1998). "It could be," Lloyd contends, "that oxytocin release upon orgasm is simply a side effect of the important role of uterine contractions during labor." She then claims that even if the uterine contractions could be related to orgasm and fertility, "there would still be questions about whether [this effect] increased reproductive success. Moreover, even if a contribution to current reproductive success could be shown, it would be insufficient to conclude that the trait evolved as an adaptation." Yet Lloyd does not entirely close the book on the possibility of a functional significance for women's orgasm, for she does acknowledge (albeit seemingly grudgingly) that "female orgasm may very well turn out to be an adaptation, exquisitely designed for some special but obscure function."

We contend that Lloyd overlooks important considerations that argue against her conclusions. Thus, while dismissing the significance of women's orgasm, she accepts men's orgasm as necessary for reproduction. She conflates men's ejaculation with men's orgasm. However, there is a big difference between the *physical act* of ejaculation and the *feeling* of orgasm. While seminal emission and ejaculation are essential to pregnancy, the feeling of orgasm is not. There is no inherent imperative that the feeling of orgasm must be linked to ejaculation. Indeed, ejaculation of viable, pregnancy-producing sperm can occur in men with spinal cord injury who do not experience the feeling of orgasm.

No adaptive function has actually been *demonstrated* for men's orgasm. Consequently, contrary to Lloyd's contention, there is no better adaptational explanation for the existence of men's orgasm than for the existence of women's. By the same token, just because adaptive significance has not been *demonstrated* for women's *or* men's orgasm, one is not justified in concluding that none exists.

We maintain, as proposed elsewhere (Komisaruk & Whipple, 1995, 1998, 2000), that women *and* men derive pleasure from orgasm and it is the pleasure of orgasm that helps to reinforce the performance of sexual intercourse, thereby promoting procreation.

It also seems plausible that even if not essential to *pregnancy,* the comparative degree of a woman's pleasurable feeling of orgasm elicited by penile stimulation during sexual intercourse with one man compared with another could be a significant factor in her selection of a mate. Similarly, the comparative degree of orgasmic pleasure that a man experiences with one woman compared with another could be a significant factor in his selection of a mate. In addition, the overt behavioral excitement—muscle tension, movement, vocalization—expressed by a woman during her sexual intercourse–induced orgasms could be a significant factor in a man's preference for that woman as a mate, and vice versa for the woman's preference for that man as a mate.

In other words, while the adaptive significance of women's orgasm has not been *demonstrated,* neither has that of men's, but this is not justifiably a criterion by which to conclude that women's orgasm is any less adaptive than men's. Perhaps a test of the adaptive significance of women's or men's orgasm is limited only by scientists' imagination and research.

Lloyd contends that women's orgasm that results from stimulation of the clitoris is a "by-product" of men's penis stimulation–induced orgasm. She points out, correctly, that the clitoris receives sensation from the pudendal nerve, the same nerve that provides penile sensation, and that this nerve conveys orgasm-producing stimulation from the clitoris in women and the penis in men. However, Lloyd ignores the extensive literature that demonstrates the existence of three *additional* pairs of nerves in women that convey sensation to the brain from the vagina, cervix, and uterus. There is good evidence that activation of these nerves by physical stimulation of these uniquely female organs can generate orgasm. As described above, these three pairs of nerves are the pelvic (sensory primarily from the vagina and cervix), hypogastric (sensory primarily from the cervix and uterus), and vagus (sensory primarily from the region of the cervix and uterus). Several studies provide evidence that direct mechanical stimulation of the vagina or cervix in the *absence* of direct clitoral stimulation can generate orgasms in women (Komisaruk & Whipple, 1994; Whipple, Gerdes & Komisaruk, 1996; Komisaruk, Gerdes & Whipple, 1997; Komisaruk, Whipple, Crawford, et al., 2004). Furthermore, women describe the *quality* of orgasms resulting from vaginal stimulation—"deep, heaving," for example—as feeling different from orgasms induced by only clitoral stimulation (J. Singer & I. Singer, 1972; Ladas, Whipple & Perry, 1982, 2005).

Thus, women have a unique and richly developed pattern of sensation that can be stimulated physically by penile intromission during sexual intercourse. (Indeed, the commercial market for dildos is testimony to the pleasure-producing nature of vaginal sensation.) Women's orgasms, we argue, are generated by their own unique organs and neural systems; they are far more than just a by-product or side effect of men's orgasms.

Addressing a physiological correlate of women's orgasm, Lloyd claims that "it could be that oxytocin release upon orgasm is simply a side effect of the important role of uterine contractions during labor." And even if the peristaltic contractions could be related to orgasm and fertility, "there would still be questions about whether [oxytocin release] increased reproductive success. Moreover, even if a contribution to current reproductive success could be shown, it would be insufficient to conclude that the trait evolved as an adaptation."

These are stringent demands that raise the question of how many other physiological processes in men *or* women that are believed to play a role in reproduction could satisfy Lloyd's proposed criteria. Much of the research on reproductive physiology is inductive, in the sense that any observed phenomenon is perceived as having a likely function. An example of this is the experimental evidence that oxytocin, released in peak levels during orgasm in women, produces a dramatic increase in uterine peristaltic movement, which propels sperm selectively into the particular fallopian tube (left or right) that receives the ovum during that particular menstrual cycle. Women in whom the greater number of sperm end up in the fallopian tube that receives the mature ovum for that cycle have a greater probability of becoming pregnant than women in whom the other fallopian tube receives the greater number of sperm (Wildt et al., 1998). Thus, by inductive reasoning, it seems likely that the oxytocin released during orgasm in women could promote this selective distribution of sperm. Although the adaptive significance has not been explicitly *demonstrated,* we can reasonably conclude that orgasmic oxytocin in women has a plausible adaptive physiological role in promoting reproduction. Much of reproduction research is inductive in the same way.

In summary, then, the absence of evidence to support an adaptive role for a physiological process in reproduction should not be taken as evidence of the absence of such a role. And the absence of evidence of an adaptive role for women's orgasm should not be taken as evidence that it is "simply" a by-product or "side effect" of men's orgasm and that it has no role. It is more productive to take the perspective that because women's orgasm has properties and underlying mechanisms that are unique (i.e., different from men's orgasm), it is likely to possess unique functions. The greater challenge is to design experiments that test hypotheses on the possible functions of women's orgasm.

3

Bodily Changes at Orgasm

Many aspects of the physiological mechanisms that are coordinated with orgasm remain unsolved, but we do know that the physiological, behavioral, and perceptual events in women's and men's orgasms comprise an intricately orchestrated and reciprocal process that is remarkable in its complexity. From specialized adaptations at the molecular level to complex social interaction, a fascinating range of entities and actions underlies both orgasm and the reproductive process. Here, we present some insights into the process of orgasm by examining several of the components and circumstances that surround the event.

Male Erection

Sexual stimulation of the penis or psychogenic (cognitively induced) arousal activates parasympathetic nerves extending from the sacral (pelvic) region of the spinal cord and terminating in the corpora cavernosa of the penis—the tissues that become engorged with blood to produce erection. (The parasympathetic system is a component of the involuntary, or "autonomic," nervous system.) There is substantial evidence that these sacral parasympathetic nerves are stimulated by oxytocin-containing neurons that originate in the brain—in the paraventricular nucleus of the hypothalamus—and project their axons, the nerve fibers that communicate with other neurons, down the spinal cord as far as the sacral level. Indeed, Argiolas and Melis (2005) reported that electrical stimulation of the paraventricular nucleus in male rats elicits penile erection and that oxytocinergic neurons (neurons that communicate with other neurons by releasing oxytocin as their chemical transmitter) are of particular importance in penile erection. These neurons project to the lumbar-sacral region of the spinal cord (the region that controls the lower back and pelvic region) to synapse with neurons that course in the hypogastric and pelvic nerves and in the sympathetic chain (an autonomic nerve pathway) to the pelvic plexus, and from this through the cavernous nerves directly to the corpora cavernosa of the penis. These oxytocinergic neurons are seemingly the only descending pathways by which the central nervous system influences erectile function (at least, the only ones identified with any degree of certainty to date).

This oxytocinergic spinal cord projection pathway is unusual because in this pathway oxytocin is acting as a neurotransmitter, although it also has hormonal functions. In milk ejection in women, for example, the oxytocinergic neurons in the paraventricular nucleus release oxytocin into the bloodstream from the posterior pituitary gland. Oxytocin is also released into the bloodstream during orgasm in both women and men (Carmichael, Warburton, et al., 1994). Therefore, oxytocin—a "neurohormone"—can act both as a hormone, by its release into the bloodstream, from which it stimulates involuntary muscle contraction of the breast and uterus, and as a neurotransmitter, by its release within the central nervous system (spinal cord and brain), where it stimulates other neurons.

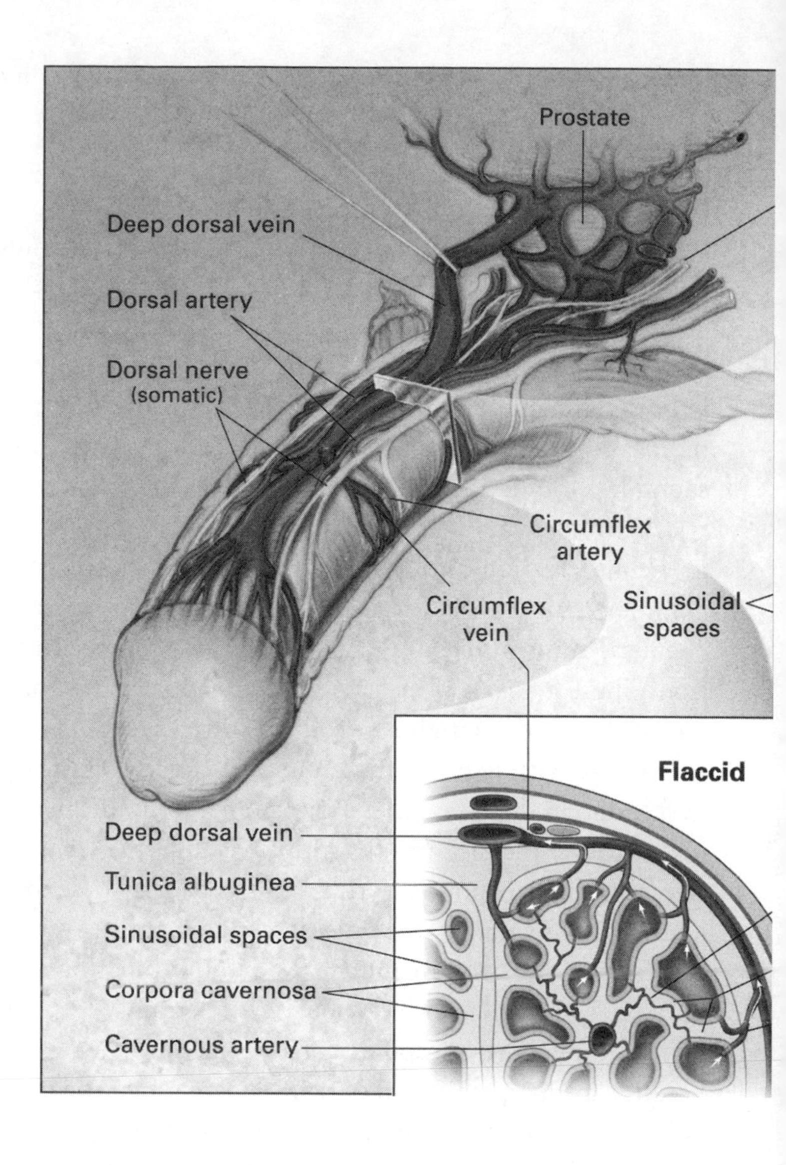

Oxytocin also functions as a neurotransmitter in the spinal cord in females, to dilate the pupils of the eyes. We do not understand the function, if any, of this pupil dilation, but it is a very distinct response in women during vaginal self-stimulation leading to orgasm (Whipple, Ogden & Komisaruk, 1992). Pupil dilation is known to be an indicator of strong arousal and interest in humans. We also know that in rats, the female's pupils dilate dramatically when the male rat ejaculates into her during mating (Szechtman, Adler &

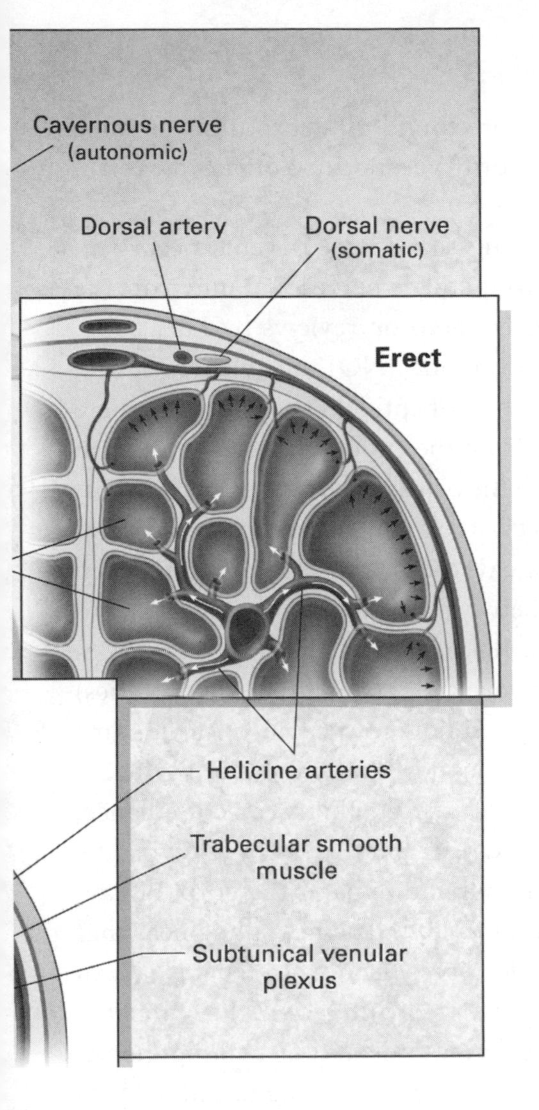

Stimulation of the skin of the penis sends impulses along the dorsal nerve to the spinal cord, reflexively activating the cavernous nerve, which relaxes the trabecular smooth muscle of the penis. This allows the sinusoids to dilate and the helicine arteries to expand, which increases the blood flow into the sinusoidal spaces of the corpora cavernosa, thus filling them. The filling sinusoids press outward against the nonelastic capsule (tunica albuginea), compressing the penile veins against the tunica and preventing blood from leaving the penis. The increased flow of blood into the penis, and the decreased escape of blood, creates the erection. The main nerves and blood vessels that supply the penis are labeled. (Image courtesy of the *New England Journal of Medicine;* from Lue, 2000)

Komisaruk, 1985). This response is due to the vaginal stimulation of oxytocin release into the spinal cord at the site of origin of the nerves that control the pupils. Several pieces of evidence in laboratory rats support this: (1) vaginal stimulation releases oxytocin in measurable amounts into the spinal cord (Komisaruk & Sansone, 2003); (2) injection of an oxytocin receptor–blocking drug into the spinal cord at that (thoracic) site attenuates the pupil-dilating effect of the vaginal stimulation (Sansone & Komisaruk, 2001);

and (3) oxytocin receptor sites are found on neurons in the same (thoracic) region of the spinal cord (Veronneau-Longueville et al., 1999).

In human males, the following sequence of events occurs on stimulation of the sacral parasympathetic nerves that innervate the penis (this information is largely based on reviews by Lue, 2000, 2001). The nerves release nitric oxide (NO), acetylcholine, and vasoactive intestinal peptide (a neuropeptide). They also stimulate the release of relaxing factors from the endothelial cells that form the lining of the blood vessels in the penis. The released neurotransmitters have the combined effect of relaxing the smooth ("involuntary") muscles of the corpora cavernosa, the corpus spongiosum (the column of erectile tissue between the corpora cavernosa), and the blood vessels of the penis. Relaxation of the smooth muscles that control the diameter of the arterioles (small-diameter arteries) to the penis allows increased blood flow to the penis. At the same time, relaxation of the smooth muscles (known as "trabeculae") of the corpora cavernosa increases the flexibility, or compliance, of the cavernosal sinusoids (blood reservoirs), thereby allowing the increased blood flow to rapidly fill and expand the corpora cavernosa. The connective tissue capsule, the tunica albuginea, that encloses the corpora cavernosa is not resilient or stretchable. When the corpora cavernosa become engorged with blood, they "inflate" against the nonresilient encapsulating tunica, thus making the penis rigid. The compression against the tunica albuginea almost totally closes down the veins that drain blood from the penis, thus trapping blood within the corpora cavernosa. At full erection, the pressure builds to approximately 100 millimeters of mercury (mm Hg), which is somewhat higher than a typical diastolic blood pressure (70 to 90 mm Hg).

The NO, acetylcholine, and vasoactive intestinal peptide released into the penis in response to the sacral parasympathetic nerve stimulation are termed "first messengers." The NO and the vasoactive intestinal peptide stimulate the production of "second messengers," which are cyclic nucleotides. Relaxation of the smooth muscle of the blood vessels and of the corpora cavernosa is produced by the action of two cyclic nucleotides, cGMP (cyclic guanosine monophosphate) and cAMP (cyclic adenosine monophosphate). The phosphate in these compounds combines with protein molecules in the smooth muscle in a process of "protein phosphorylation," causing an expansion of the muscle protein and consequent relaxation of the muscles, much like fingers loosening their grip on a hose. The dilating blood vessels increase the blood flow into the penis, leading to erection as described above.

Because penile erection is so dependent on adequate blood flow, erectile dysfunction can provide an early warning "red flag" for impending vascular (blood vessel) disease. Erectile dysfunction in otherwise asymptomatic men may be a marker of a silent vascular disease, especially coronary artery disease, and provides an important new means of identifying those at risk for such disease (Kirby et al., 2001).

Erection is normally short-lived, because an enzyme in the penis, phosphodiesterase-5 (the suffix *-ase* denotes "enzyme"), breaks down the cAMP and cGMP, thereby inactivating them. Pharmacological blocking of the phosphodiesterase activity prolongs the action of the two nucleotides, intensifying and prolonging the smooth muscle relaxation and hence the erection. This is the mechanism of action of sildenafil and chemically related erection enhancers, known by the trade names Viagra, Cialis, and Levitra. An interesting historical note is that this erection-enhancing effect was discovered serendipitously in the course of developing a blood pressure–lowering drug for the treatment of angina pectoris (cardiac-related constricting pain in the chest).

Because the erection-enhancing drugs require the presence of the cyclic nucleotides in order to act, their effect requires stimulation of the nerves that release the cyclic nucleotides in the penis. This nerve stimulation is produced either reflexively by mechanical stimulation of the penis or by cognitive sexual arousal. Reflexively,

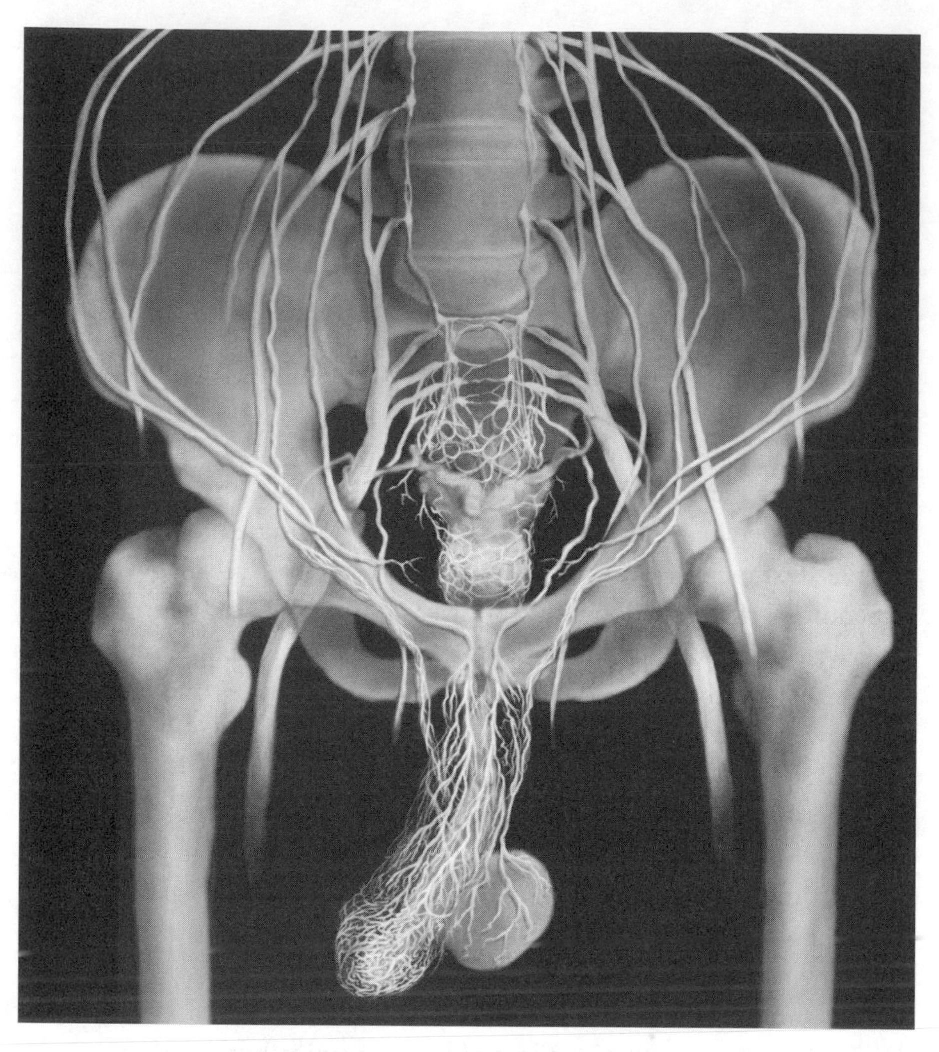

The nerve supply of the male genital system. Note the dense nerve supply especially to the penis, the testicle, and the prostate gland, which is located centrally inside the pelvis. (Image courtesy of Anatomical Travelogue)

the sensory signals are transmitted into the spinal cord, where they stimulate output via the nerves projecting to the penis. Cognitively, the brain sends neural signals down the spinal cord and out via the same nerves to the penis.

In addition to the nerves that control the smooth, or "involuntary," muscles of the penis, another pair of nerves, the pudendal nerves, convey impulses to the "voluntary," striated muscles of the perineum—the ischiocavernosus muscles. Contraction of these muscles intensifies the erection. When the ischiocavernosus muscles are contracted voluntarily, they compress the base of the blood-filled corpora cavernosa and the erection becomes more rigid. During this phase, both the inflow and outflow of blood temporarily cease, and the intracavernous pressure may reach a magnitude of *several hundred* millimeters of mercury, much higher than resting blood pressure.

Contraction of the ischiocavernosus muscles also moves the erect penis to a more upright position and participates in the rhythmical expulsion of semen during ejaculation. In addition, voluntary contraction of these muscles has the effect of contracting the external anal sphincter. These are the same muscles that women and men contract voluntarily in Kegel exercises, used by people wishing to control urinary incontinence.

The detumescence of (emptying of blood from) the penis that occurs after ejaculation is produced by activation of the involuntary-sympathetic hypogastric nerves. Detumescence can result from cessation of neurotransmitter release, breakdown of second messengers by phosphodiesterases, or sympathetic nerve discharge, or any combination of these. Contraction of the trabecular smooth muscle reopens the venous channels, the trapped blood is expelled, and flaccidity (loss of erection) returns (Lue, 2000, 2001).

The sympathetic nerves are apparently tonically active (*tonic* describes a continuous or sustained level of activity), since intracavernous administration of drugs that block the action of the neurotransmitters released by the sympathetic nerves—drugs such as phentolamine or phenoxybenzamine, known as "alpha-adrenergic receptor blockers"—produces erection in men. Brindley (1986) demonstrated the potency of this mechanism:

> In the course of developing treatments for erectile impotence and priapism [prolonged erection] . . . I made, on myself as subject, observa-

tions on the actions of a number of other drugs . . . The very prolonged erection caused by phenoxybenzamine and phentolamine . . . [is] entirely undiminished by concentration on exacting intellectual tasks . . . Objective indications that psychological factors contribute little . . . are the fact that during one phentolamine erection I took an urgent and worrying telephone call without losing the erection. On the other hand, attention to involuntary swelling of the penis is somewhat sexually arousing, so a slight or moderate tumidity caused by a drug is almost certainly sometimes increased briefly by the act of measuring it.

Male Orgasm and Ejaculation

In a review published in 2005, Levin stated that "remarkably, a detailed, non-disputed physio-anatomical description of the mechanism of human ejaculation has still to be produced." Two main questions have evidently not been resolved. How exactly are the spurts of semen produced? And, is the pleasurable sensation of orgasm a consequence of the sensory activity produced by the ejaculation mechanics, or is it primarily a brain phenomenon that is additively and pleasantly enhanced by the sensory activity generated by the expulsive flow of semen at ejaculation? We review here some of the most relevant experimental evidence and present a presumptive scenario, with the disclaimer that it will certainly need revision as more information becomes available. This is a case in which an understanding of the *structure* of a system is key to understanding its *function.*

From a purely mechanical standpoint, smooth muscle contraction alone is probably not strong enough or sudden enough to emit semen in spurts. Therefore, precise temporal coordination is necessary between the smooth muscle contraction of the ejaculatory duct system and of the sphincter muscles (acting much like purse strings) interposed in the path of semen flow. Thus, (1) the ducts fill (in the process of emission), (2) pressure builds up behind a closed sphincter valve (the entity consisting of the semen-filled duct and

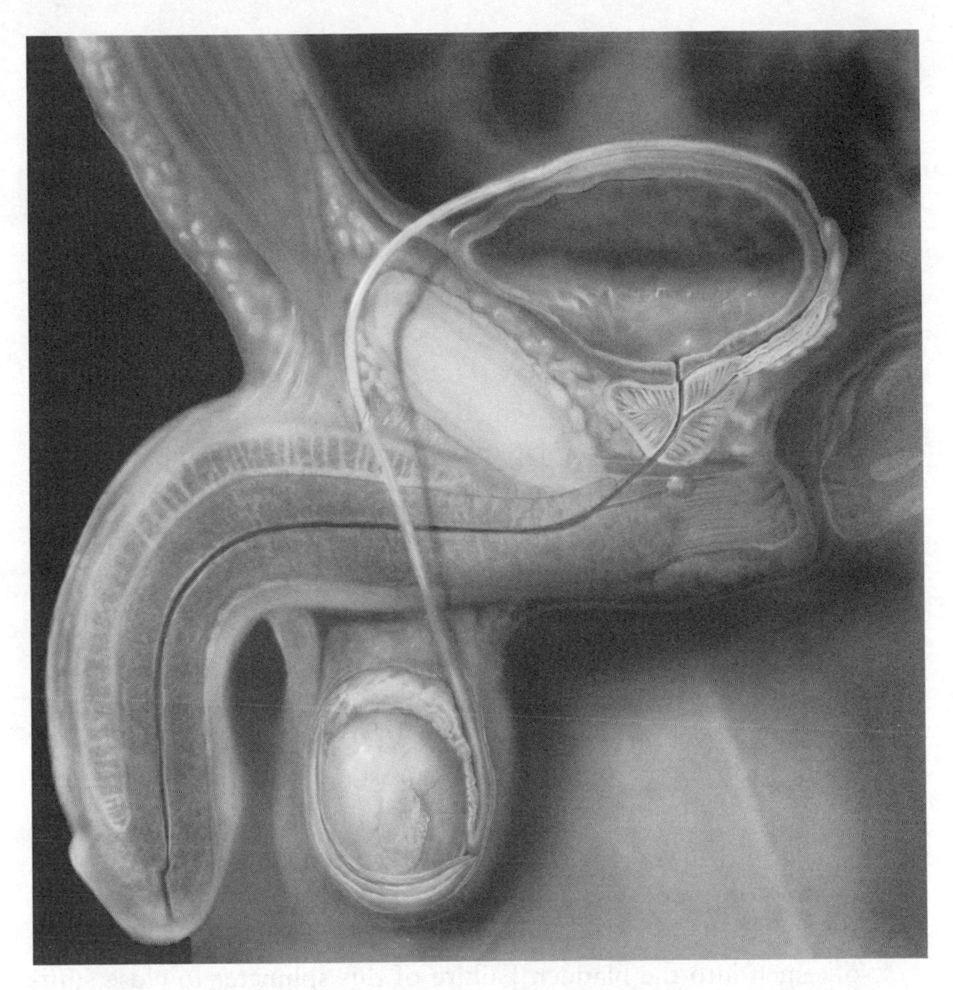

The anatomy of the male genital system. The sperm begin their journey by leaving the testicle via the thin, tubular vas deferens, which passes up and over the large, oval-shaped (light colored) pubic bone, then over (horizontally across the top right of the image), behind, and under the urinary bladder (large hollow oval). At this point the sperm pass through the prostate gland (below the bladder and to the right of the pubic bone), then immediately enter the urethra (thin, dark-colored passageway). Seminal fluid is added into the urethra from the prostate and the seminal vesicles (elongated structure behind [to the right of] the urinary bladder). The sperm and seminal fluid pass along and out through the penis. (Image courtesy of Anatomical Travelogue)

sphincter is termed the "pressure chamber"), (3) sudden release of the sphincter valve emits a spurt of semen (ejaculation), (4) the ducts fill again behind a reclosed sphincter, and (5) a repeat of the process results in another spurt of semen. The process repeats several times in rapid succession in a matter of a few seconds.

The initial, emission, stage in the ejaculation process is produced by contractions of the smooth muscle in the capsules of the testes, seminal vesicles, and prostate and the smooth muscle of the ducts of the epididymis and vas deferens. The vas deferens contracts as a unit, from the caudal (end part) epididymis to the ejaculatory duct of the prostate; the seminal vesicles contract by peristalsis (a traveling wave of muscle contraction and relaxation in this tubular organ that propels the semen onward). The contractions of this smooth muscle are stimulated by activation of the sympathetic division of the autonomic nervous system via the hypogastric nerve (which exits the spinal cord at levels thoracic 10 to lumbar 2). The evidence for this mechanism came from the observation that administration of a sympathetic division receptor blocker (phentolamine or phenoxybenzamine) prevented discharge of semen into the urethra (Brindley, 1986; Gerstenberg, Levin & Wagner, 1990). Also, electrical stimulation of the hypogastric nerves produced contraction of the bladder neck, prostate, seminal vesicles, and ejaculatory ducts, thereby simulating natural emission.

The pressurized semen is retained in the duct system by contraction of sphincters. Contraction of the proximal sphincter (also known as the internal prostatic sphincter) prevents the backflow of semen into the bladder. Failure of this sphincter to close sufficiently during ejaculation permits semen to make a U-turn and enter the urinary bladder, which is termed "retrograde ejaculation." Concurrently with contraction of the proximal sphincter, the distal sphincter (the external prostatic sphincter) relaxes and prostatic contractions begin (Vale, 1999). This allows the pressurized semen to move from the "bulbous" urethra, which is in the pelvis, into the "pendulous" urethra, which is in the penis. This sphincter coordination allows typical, anterograde (forward) ejaculation.

At orgasm, under the control of the pudendal nerve (which exits the spinal cord at sacral levels 2 to 4), the pelvic striated muscles, especially the bulbocavernosus muscle, start contracting rhythmically and involuntarily. Brindley (1986) pointed out a curious property of this muscle: "The contractions of the bulbocavernosus muscles, though they are of voluntary muscle, are strictly reflex and not inhibitable by any effort of will." Furthermore, while involuntary contraction of this muscle during ejaculation feels pleasurable, voluntary contraction of the muscle at other times does not produce a comparably pleasurable sensation (Levin, 2005).

The distal sphincter opens and closes in closely coordinated temporal relation with the contraction of the bulbocavernosus muscle. The intermittent opening and closing aids in propulsion of the semen through the bulbous urethra to the pendulous urethra. The importance of the coordination of the opening and closing of the distal sphincter is shown by the finding that if the sphincter becomes paralyzed or is incapacitated by experimental anesthesia, a dripping seminal emission, rather than ejaculation, occurs (Levin, 2005).

Also contributing to ejaculation is the contraction of the perineal muscles and anal sphincter, mediated by the voluntary nervous system. In addition, bladder neck contraction occurs synchronously under the control of the sympathetic division of the involuntary (autonomic) nervous system (Vale, 1999).

As pointed out by Levin (2005), the existence of a pressure chamber created in part by the action of the sphincters, a concept initially proposed by Marberger (1974), has not been confirmed. Two sonographic studies failed to find evidence of the ballooning of the urethra that would support the pressure chamber concept (Gil-Vernet et al., 1994; Hermabessiere, Guy & Boiteaux, 1999). However, it is possible that the methodology did not have sufficient resolution to detect the expected distention of the duct system. Thus, despite lack of positive evidence, it would be premature to dismiss the pressure chamber concept of ejaculation described above.

There is still the unresolved issue of what initially triggers the cascade of events leading to ejaculation. Is it triggered reflexively by the sensory activity generated by the increased pressure of semen in the duct system? Or is it triggered independent of the pressure in the duct system, when excitation in the brain or spinal cord reaches a critical level? The evidence leans toward the latter.

> The major experimental evidence against the involvement of semen volume specifically initiating the striated muscle contractions [of orgasm] is that subjects taking [sympathetic nerve] adrenergic blockers have striated muscle contractions with orgasm, but little or no ejaculate is produced because the emission phase is blocked. Our two subjects who were given phenoxybenzamine were both able to stimulate themselves to orgasm and the bulbocavernosus muscles fired off in their normal clonic pattern [i.e., rapid alternation of muscle contraction and relaxation]. No semen was ejaculated in one of the subjects while the other had at least 3 contractions of the bulbocavernosus muscle before the reduced semen volume was ejaculated . . . Both subjects reported that their orgasms felt normal and were apparently uninfluenced by the adrenergic blocker. It is unlikely that the trigger for the initiation of the bulbocavernosus muscle contractions is dependent on a critical volume distension by the semen of any internal organs. All we can postulate is that central/spinal ejaculation centres activate the contractions when they reach a critical firing level. (Gerstenberg, Levin & Wagner, 1990)

Shafik (1998) provides further evidence against the involvement of semen pressure in the urethra in triggering ejaculation. He inserted an expandable bulb into different regions of the urethra of human subjects and found that the bulbourethral striated muscle contraction that accompanies ejaculation was not elicited by a normal magnitude of distention of the urethra, but only by supernormal distention. According to Levin (2005), "What is certain is there are definite occasions where the ejaculatory mechanism is activated, yet no seminal fluids enter the prostatic urethra."

On the question of where orgasm is perceived, certainly the flow of semen through the urethra can be perceived and is pleasurable.

However, it probably *contributes to* the pleasurable sensation of orgasm rather than creating it, as evidenced by the existence of dry orgasms. Thus, the trigger for ejaculation does not seem to be a peripheral stimulus but is probably generated by the brain and conveyed to the spinal cord and then out to the periphery.

Two other curious aspects of orgasm in men are noteworthy. Brindley (1983) noted that "ejaculation and orgasm do not necessarily require erection. Erection can be prevented by compression of the pudendal arteries by a mechanical device. If this is done, ejaculation and orgasm can still occur, the penis remaining entirely flaccid." And in rare cases, orgasm in men can occur without any mechanical stimulation. According to Kinsey, Pomeroy, and Martin (1948), three or four men out of five thousand reported being able to experience ejaculatory orgasm without mechanical stimulation.

Spinal Cord Control of Erection and Ejaculation

Erection and ejaculation are under the direct control of specific parts of the spinal cord, which in turn are controlled by the brain. Complete spinal cord injury, producing an absence of sensation and voluntary movement below the level of the injury, is devastating to people's lives. Some men and women with such injury have told us that the most troublesome aspect is not the inability to walk or even compromise of their sexual activity but their loss of control over bladder and especially bowel function.

The effects of spinal cord injury on erection and ejaculation depend to a significant extent on the location and completeness of the injury. The genital outflow nerves leave the spinal cord at two levels: one set of nerves (sympathetic nerves) at the lumbar level, just below the midlevel of the ribs, and the other set (parasympathetic nerves) at the sacral level, near the tailbone.

If the spinal cord injury is above the midrib level (closer to the head), the impulses from the brain that descend through the spinal cord to both sets of genital nerves are blocked, preventing penile erection in response to sexual thoughts ("psychogenic" erection). However, because these nerves carrying impulses between the spinal cord and the genitals are not themselves damaged by the injury, erection can be elicited reflexively (via the parasympathetic nerves) by direct mechanical stimulation of the penis. But because the impulses produced by the direct penile stimulation cannot reach the brain, the person cannot feel the stimulation.

With the same, above-midrib level of spinal cord injury, direct penile stimulation can also reflexively elicit ejaculation (via the sympathetic nerves). But, if the injury is lower down the spinal cord (between the lower-rib level and the tailbone), direct penile stimulation can produce reflexive erection but not ejaculation, because although the input from the penis to the parasympathetic nerves is intact, it cannot ascend the spinal cord to reach the sympathetic nerves that control ejaculation. In such cases, because the person cannot feel penile stimulation, he reportedly does not have an orgasm. (For details, see Bors & Comarr, 1960; Chapelle, Durand & Lacert, 1980; Netter, 1986; Shafik, 1998, 2000; Steers, 2000; Bird et al., 2001; Coolen et al., 2004; McKenna, 2005.)

Female Orgasm and Reproduction

During orgasm in women the uterus contracts rhythmically. A controversial concept of the role of these contractions is that they have an "insuck" effect that pulls semen into the uterus, from where sperm cells pass up the fallopian tubes and meet the ovulated egg. If fertilized, the ovum passes down into the uterus and implants into the uterine wall.

At orgasm, a woman's blood pressure and heart rate reach approximately double their resting levels. This increases the rate of blood flow in the body and thus carries increased oxygen and nutrient supply to the muscles and other organs, supporting their increased metabolic activity. There is also a marked reduction in sensitivity to pain at orgasm, to about half the resting level, although, curiously, sensitivity to touch is not decreased and may even be enhanced (Komisaruk & Whipple, 1984; Whipple & Komisaruk, 1985). Anecdotally, it has been noted that intense muscular activity at orgasm may lead to bruising or scratching without awareness of pain, yet simultaneously, a single hair on the tongue may be particularly distracting. This indicates how the different sense modalities (touch and pain) are differentially affected during orgasm. The insensitivity to pain may promote a more intense forcefulness of bodily contact and intromission than would be acceptable to the partners otherwise, which could lead to a reflexive intensification of the uterine contractions, thus promoting the insuck of semen into the uterus. The timing of the insuck of semen in relation to sperm capacitation (the process that renders sperm capable of fertilizing an egg) has been discussed critically by Levin (2002).

The original model of sexual response in women—that is, the bodily events leading up to and including orgasm—was proposed by Masters and Johnson (1966):

—Excitement phase: image, fantasy, memory, sensory stimulus
—Plateau phase: increased blood flow to vagina, vaginal dilation, vaginal lubrication
—Response or orgasm phase: muscle contraction of lower anus and lower pelvic muscles, uterine contractions, vaginal contractions (orgasm)
—Resolution phase: satisfaction, relaxation (decreased blood flow to sex organs, relaxation of muscle tone), euphoria

This characterization was later modified and expanded to view the plateau phase not as a separate stage but as a late-stage excitation (Robinson, 1976), and then to add an earlier stage (Kaplan, 1979), "desire," as first characterized by Lief (1977). The concept of sexual desire was subsequently differentiated into two different types: spontaneous or induced by physical excitation (Levin, 2002).

Giuliano, Rampin, and Allard (2002) recently expanded the

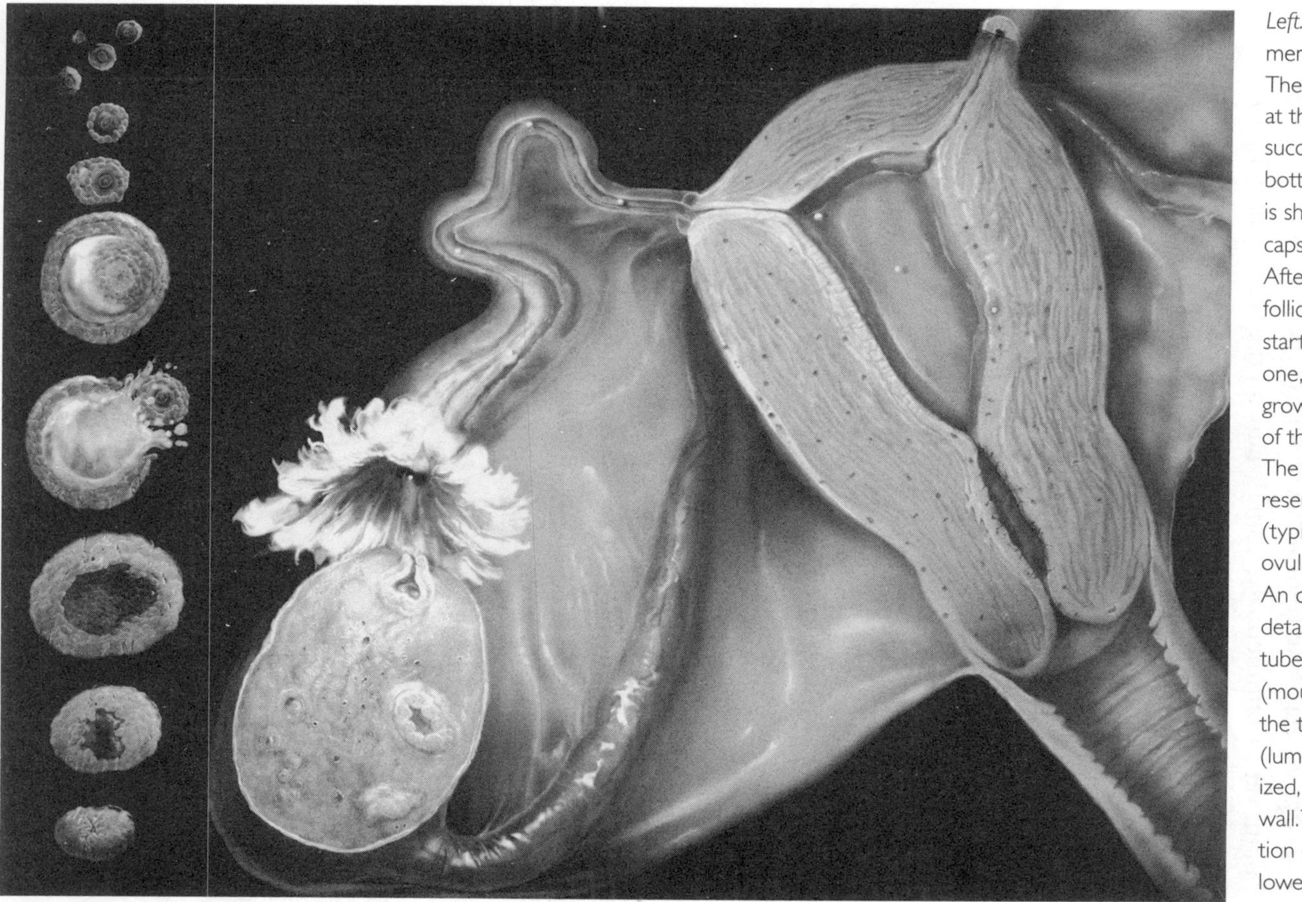

Left. Stages in the development of a human egg (ovum). The earliest stages are shown at the top of the image, with successive stages toward the bottom. In the middle, the ovum is shown bursting out of its capsule (the ovarian follicle). After the ovum is released, the follicle becomes smaller and starts to secrete progesterone, which stimulates uterine growth. *Right.* The anatomy of the female genital system. The path of the ovum is represented here by multiple ova (typically, just one ovum is ovulated per menstrual cycle). An ovum (oversized to show details) enters the fallopian tube at its feathery ostium (mouth) and travels through the tube into the uterine cavity (lumen). If the ovum is fertilized, it implants in the uterine wall. The cervix and upper portion of the vagina are at the lower right side of the figure. (Images courtesy of Anatomical Travelogue)

characterization of the plateau and response phases. They describe "female genital sexual response" as a "combination of vasocongestive and neuromuscular events in the genital tract and pelvic floor which include blood engorgement of the vagina and clitoral erection . . . The genital structures that undergo the vasculogenic changes are the labia, the periurethral glans, the urethra, and Halban's fascia. The inner vaginal lumen widens first, followed by a progressive increase of vaginal luminal (tubular space) pressure from sexual arousal to orgasm. This culminates in a series of clonic [repetitive] contractions of striated ['voluntary,' e.g., pelvic floor] and smooth ['involuntary,' e.g., vaginal] muscles at orgasm."

Maravilla et al. (2005) describe a sequence of engorgement of the labia and clitoris during sexual arousal. On initial erogenous stimulation, there is increased blood flow to the internal and external genitals, including the clitoris and labia, and they become engorged with blood.

The lubrication and genital swelling that occur during sexual arousal may be reduced in women who have undergone total hysterectomy (removal of uterus and cervix) if the lateral anterior and posterior ligaments supporting the uterus and cervix are also removed. These ligaments provide the anatomical pathway for the pelvic autonomic nerves between the uterus and cervix and the spinal cord (Maas et al., 2003).

Fertilization

An intricately coordinated physiological process between male and female sexual systems enables fertilization. In women, orgasm is evidently not one of the components that is *essential* to fertilization. As Vance and Wagner (1976) stated, "There appears to be no relationship between orgasmic capacity in the female and her fecundity" (*fecundity* here means likelihood of becoming pregnant). However, several studies, reviewed below, suggest that orgasm may *assist* the process.

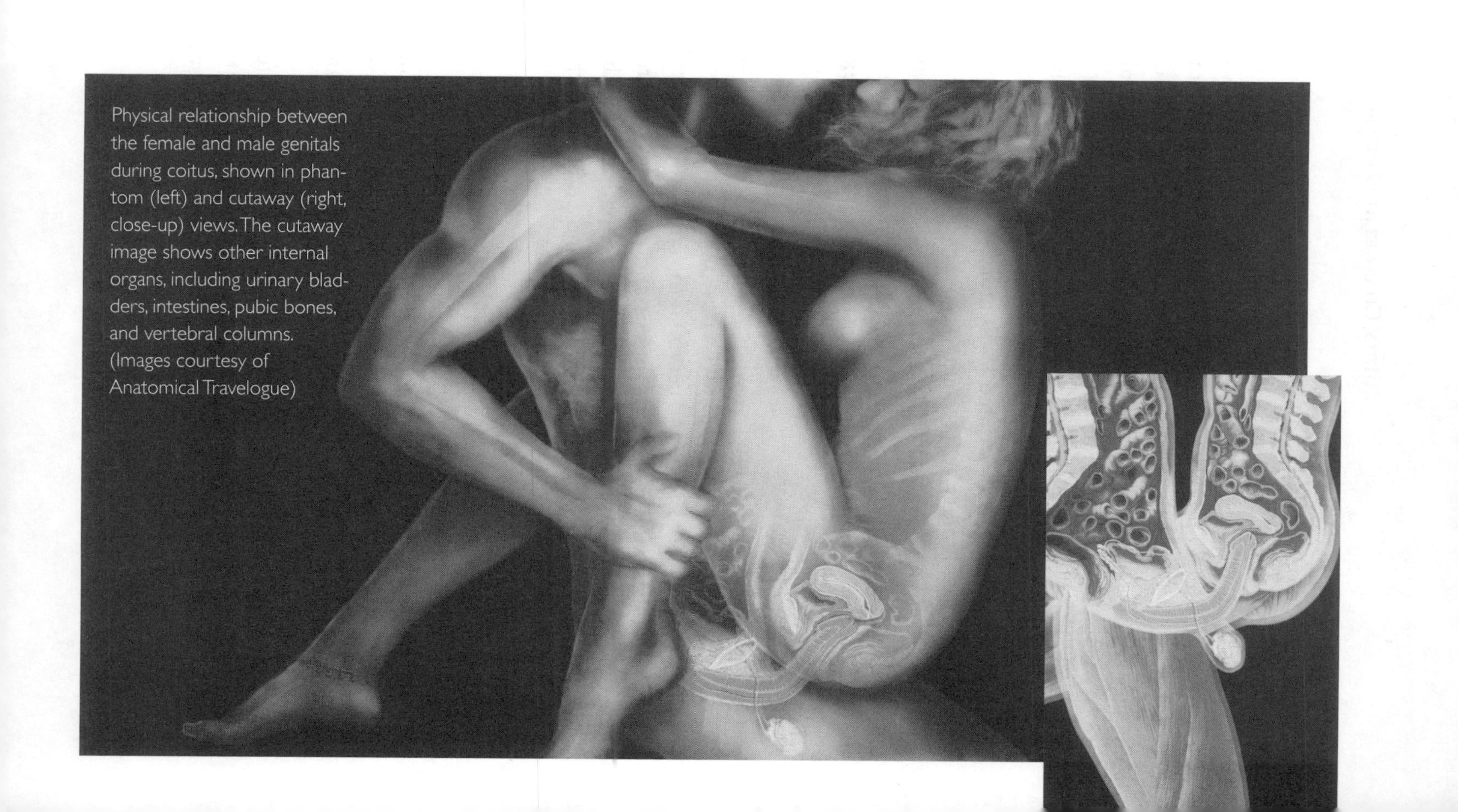

Physical relationship between the female and male genitals during coitus, shown in phantom (left) and cutaway (right, close-up) views. The cutaway image shows other internal organs, including urinary bladders, intestines, pubic bones, and vertebral columns. (Images courtesy of Anatomical Travelogue)

The cyclical growth and shrinkage of the vaginal, cervical, and uterine epithelium (the layer of cells forming the internal lining of these organs) and their blood supply is controlled by the cyclical waxing and waning of estrogens and progestins, steroid hormones secreted mainly by the ovaries.

The vaginal epithelium is not glandular—that is, does not secrete any substances—yet a vaginal lubricant is produced. In response to the release of vasoactive intestinal peptide by nerve endings in the vagina, the slippery fluid passes from the blood in the capillaries through the cells of the vaginal lining and into the vaginal canal. This process, termed "transudation," is fundamentally distinct from the more familiar lubrication systems that involve secretion by glands (Levin, 1998). Another neuropeptide, neuropeptide Y, may contribute to transudation, in addition to a relatively recently recognized neurotransmitter—nitric oxide (the same neurotransmitter that stimulates penile erection) (Hoyle et al., 1996). This neurotransmitter process also increases the blood flow and feeling of congestion in the vagina, and its buffering action neutralizes the initial acidity of the vagina, thereby promoting sperm metabolism, motility, and survival (Levin, 1998).

Some sperm cells move freely in the semen, but others are temporarily trapped in the semen coagulate. Enzymes in the semen gradually decoagulate the semen to liberate sperm, thereby pacing the release of sperm and entry into the uterus. The semen also contains prostaglandins, some of which stimulate, and others inhibit, the contraction of smooth muscles of the vagina, cervix, and uterus.

During sexual excitement, the vaginal canal relaxes and dilates, and smooth muscle in the connective tissue attached to the anterior wall of the vagina (i.e., the region closest to the surface of the abdomen) pulls on the anterior vaginal wall thus expanding the vaginal canal, which results in a "tenting" of the canal (Levin, 1998, 2002). Levin speculates that the tenting has two functions. First, it may reduce friction on the penis, delaying ejaculation and providing an opportunity for more stimulation by the thrusting penis to increase the magnitude of the vaginal tenting. Second, the tenting may lift the cervix away from the pool of semen following ejaculation, at the same time creating a reservoir for the semen. This may delay the passage of sperm through the cervix, allowing more time for decoagulation and hence one form of capacitation of the sperm. A caveat in Levin's model is that the lifting away of the cervix from the pool of semen may pertain in the case of ejaculation during intercourse in the "missionary" position, but not in the "rear-entry" or "woman-on-top" positions.

Alzate and Londono (1984) have suggested a different potential consequence of the tenting of the vagina during sexual intercourse. They found that 60 percent of their female subjects had never or almost never experienced a coital orgasm, indicating that "coitus is an inefficient method of inducing female orgasm." They accounted for this on the basis that if the woman reaches an advanced stage of sexual arousal, ballooning (i.e., tenting) of the upper part of the vagina could occur and would preclude any further contact between the glans penis and the erogenous zone if coitus is in the missionary position. However, if coitus is in the rear-entry or woman-on-top position, there is a much better chance for the penis to effectively stimulate the anterior vaginal wall (Gräfenberg, 1950), which could trigger an orgasm. Other investigators found no tenting when the anterior vaginal wall (and thus the G spot) was stimulated and the woman experienced orgasm. Under those conditions, the cervix was found to dip into the seminal pool in the vagina (Perry & Whipple, 1981, 1982; Ladas, Whipple & Perry, 1982, 2005). Thus, there are still significant and unanswered questions about the significance for fertilization of the vaginal tenting phenomenon.

Here we present a selectively simplified synthesis of the relationship between woman's orgasm and the transport of sperm to the fallopian tubes, where further sperm capacitation and fertilization occur. The main caveat is that some of the evidence is based on research that addressed isolated aspects of the process—such as processes that could account for semen flow through the uterus—and was carried out under non-orgasm conditions. Another caveat is that the evidence is based on studies with a widely variable number of experimental human subjects, ranging from many to just one.

The muscular activity of the uterus is spontaneous during the menstrual cycle, contractions sometimes occurring several times per minute. As an ovum matures in its ovarian follicle (the follicle is the capsule that encloses each ovum as it develops in the ovary) in the days just before ovulation, the nature of the uterine contractions changes. In the earlier days of the menstrual cycle, the contractions start at the region of the uterus termed the "fundus," which is closest to the ovaries, and progresses in peristaltic waves toward the cervical end. In the days immediately before ovulation, as an ovarian follicle matures and approaches the time of ovulation, the direction of the peristalsis reverses, starting at the cervical end and moving toward the fundus. This latter peristaltic movement creates a suction effect against the internal, uterine surface of the cervix that could aspirate sperm from the vagina into the uterus and move the sperm toward the fundus (Kunz et al., 1996). The fallopian tubes open at the fundus. They receive a mature ovum discharged from the ovary at one end (the ostium) and receive sperm at the other, fundal, end, the two cells meeting and fertilizing in the middle of the tube.

During sexual intercourse, physical stimulation of the clitoris, vagina, cervix, breasts, and nipples stimulates secretion of oxytocin from the posterior lobe of the pituitary gland into the bloodstream. The oxytocin is maximally released starting within one minute after orgasm (Blaicher et al., 1999). The oxytocin is conveyed through the bloodstream to the uterus, where it stimulates an intensification of the uterine muscle contractions. In the minutes leading up to orgasm, there is a buildup of pressure inside the uterus. This may actually expel some mucus out through the cervical opening (the os) and into the vagina. Then, within about one minute after orgasm, a negative pressure develops in the uterus that persists for about one to two minutes (Fox, Wolff & Baker, 1970). If the man has ejaculated, the cervical mucus expelled into the vagina can entrap sperm, and the negative pressure that has developed in the uterus has the potential to suck the sperm embedded in the mucus back through the cervix and into the uterus.

Several different studies have shown that sperm can be transported all the way from the vagina through the cervix, through the uterus, and into the fallopian tubes remarkably fast. This rapidity of transport cannot be accounted for by the swimming ability of sperm. Indeed, radioactively labeled *inanimate* particles (albumin tagged with technetium) were shown to be transported from the vaginal surface of the cervix all the way to the fallopian tubes within as little as one minute (Kunz et al., 1996). Clearly, motility is not necessary for the transport of sperm into the uterus. It is not clear to what extent the motility of the sperm participates in their travel. The motility may be more closely involved in the access of sperm to the ovum once they have been carried passively to its local environment in the fallopian tube (Eisenbach, 1995).

A study under natural conditions suggested that women's orgasm sucks ejaculated sperm into the uterus (R. R. Baker & Bellis, 1993, 1995). (We point out some problems with these data after presenting the authors' interpretation.) Baker and Bellis reported that a specific type of vaginal secretion, termed "flowback," is normally expelled by women after sexual intercourse. After intercourse during which the man ejaculates, the flowback contains sperm. In this study, the authors counted (under the microscope) the number of sperm contained in the flowback in relation to the time at which the woman's orgasm occurred relative to her partner's ejaculation. They found relatively *few* sperm cells in the flowback when the woman's orgasm started between one minute before and about forty-five minutes after her partner's ejaculation. By contrast, when the woman's orgasm occurred earlier than one minute before the man's ejaculation, there were far *more* sperm in the flowback. In other words, the woman's orgasm apparently generated a suction within the uterus that aspirated the sperm into the uterus, thus reducing the number of sperm in the flowback.

Other research found a curious correlation with this timing of orgasms: women's desire to become pregnant could be predicted by the frequency of orgasm *after* the partner's orgasm; frequency of orgasm *before* the partner's orgasm was not a significant predictor of desire for pregnancy. Desire to get pregnant was positively predicted by active sexual participation or "taking charge of the situation," independent of other predictors. Active sexual participation could enable the woman to manipulate the timing of her orgasm in relation to her partner's ejaculation (Singh et al., 1998).

In the light of these findings, it would be informative to know whether, in women who experience orgasm before the man ejaculates, the uterus loses the vagina-to-ovaries direction of peristalsis before ejaculation occurs and thus reduces the insuck of semen.

Levin (2002) suggested that the flowback losses in sperm count reported by Baker and Bellis (1993, 1995) could be due to vaginal tenting, which could provide a sperm-semen "reservoir space" for ejaculated semen to pool at the vaginal face of the cervix, "thus reducing potential 'flowback' losses." However, this would not explain why the sperm count in the flowback up to one minute before the woman's orgasm is so much *higher* than immediately thereafter, the tenting already occurring. The tenting concept does not seem to account for the sudden marked reduction in flowback loss that occurs from one minute before to forty-five minutes after the woman's orgasm. The sudden drop in sperm count is much better accounted for by orgasm insuck than by tenting.

There are two main problems with Baker and Bellis's research. First, the authors used a complex, highly derived (i.e., indirect) formula for quantifying the sperm in the flowback, rather than a straightforward tally of the numbers of sperm, making it difficult to know exactly what they measured. A second problem is that while they reported the changes in sperm measurement in relation to the time before and after ejaculation and the woman's orgasm, they did not report control data to show the comparable time course of their sperm measure in relation to ejaculation when the woman did *not* experience an orgasm. These two problems necessarily raise a caveat about the insuck concept. (For a further and detailed critique of the insuck concept, see Lloyd, 2005.)

Through the use of radioactively labeled particles designed to mimic sperm, researchers observed a remarkable process relevant to fertilization (Wildt et al., 1998). They applied radioactive particles to the vaginal surface of the cervix and, using a scanner outside the body, photographed their trajectory through the uterus and into the fallopian tubes. Rather than being distributed equally between the two fallopian tubes, the radioactive particles were transported predominantly to only one or the other tube. The tube to which they were transported showed a significant tendency to be on the same side as the ovary that was soon to ovulate. In other words, if the ovary on the right side had the mature, predominant follicle (the follicle containing the egg whose ovulation was most imminent), then the predominant transport of the sperm-mimics was to the fallopian tube on the right side.

When the mature follicle ruptures, it releases the ovum into the fallopian tube, and, as noted above, it is typically in the fallopian tube that a sperm encounters and fertilizes the ovum. Thus, Wildt et al.'s elegant study shows that a mechanism exists to physically maximize the access of the ovum to sperm. The authors suggested that the basis for this left-right selective transport is that the developing ovarian follicle secretes a hormone (perhaps progesterone) that produces a local relaxation of the smooth muscle of the fallopian tube nearest to the follicle, thus opening it up, while leaving closed the fallopian tube on the other side. This other side might be comparably involved during the next menstrual cycle. Wildt et al. summarized the importance of this one-sided favoritism as follows: "That ipsilateral [same-side] transport has biological significance is suggested by the observation that the pregnancy rate following spontaneous intercourse or insemination was significantly higher in those women in whom ipsilateral transport could be demonstrated than in those who failed to exhibit [such] lateralization."

Capacitation of the sperm can occur in the isthmus of the oviducts, the junction between the uterus and each fallopian tube. Capacitation is temporally coordinated with passage of the ovum from the ovaries to the fallopian tubes, thereby optimizing the timing of fertilization.

After the ovum is released from its follicle, it can survive in the fallopian tube and become fertilized for a period of about twenty-four hours. In humans, there is no precise synchrony between the time when an ovum is released from the ovary and the time of sexual intercourse. By contrast, in certain species of mammals—such as rabbits, cats, and the 13-lined ground squirrel—the female releases ova only in response to copulatory vaginal stimulation, in a process termed "reflex ovulation." This precisely synchronizes the time of ovulation with the passage of sperm. In women, by contrast, the process of ovulation is characterized as "spontaneous ovulation" and results from the spontaneously changing levels of hormones during the menstrual cycle, independent of whether or when coital stimulation occurs.

Although the released ovum has a lifespan of about twenty-four hours, sperm cells have a lifespan of only one to four hours after capacitation. To compensate for this difference, sperm are capacitated and released from the ejaculated semen pool only gradually, thereby spreading out in time the opportunity for a capacitated sperm cell to reach and fertilize the ovum: 2 to 14 percent of the sperm become capacitated at a time. This process increases the window of opportunity for fertilization between the time of ovulation and intercourse (Singh et al., 1998).

While it is widely recognized that a woman's orgasm is not *essential* to pregnancy, the studies described above provide evidence that a woman's orgasm may bring into play a combination of physical processes that *promote* pregnancy. These processes could make a critical difference under compromised physiological conditions, such as a relatively low sperm count. Furthermore, and as pointed out by several authors (e.g., Thornhill, Gangestad & Comer, 1995), the occurrence (or non-occurrence) and *timing* of a woman's orgasm, and the particular male partner, might influence to some degree the probability of her becoming pregnant.

Menstrual Cycle and Rhythm of Intercourse

Steroid hormones that are produced and secreted by the ovaries and adrenal cortex play an obligatory (essential) role in female sexual behavior in many species of animals. That is, the females become completely unreceptive to mating if the sources of the hormones are surgically removed. By contrast, these "sex steroids" do not play an obligatory role in humans. Rather, they play a modulatory role: while sex steroids do exert an influence on sexual response, women and men are, to a great extent, liberated from dependence on the sex steroids for their sexual response and can experience sexual responses despite wide fluctuations in, or absence of, these hormones.

What is the difference between an obligatory role and a modulatory role? As a humble and culinary example, an egg plays an obligatory role in an omelet but a modulatory role in a potato salad.

Estrogens stimulate growth of the vaginal epithelium. After menopause, reduction in estrogens leads to thinning of the vaginal wall, less vaginal lubrication, and a reduction of the vaginal lubrication induced by vasoactive intestinal peptide (Ottesen et al., 1987).

Statistically, there tend to be peaks of intercourse in humans around the time of ovulation, and other peaks shortly before and after menstruation. While the periovulatory peak has obvious adaptive function, the basis for perimenstrual peaks is less obvious. Some authors have suggested several explanations for the perimenstrual peak: pent-up desire after voluntary abstinence during the ovulatory peak, loss of the fear of an undesired pregnancy, or, as suggested by Levin (2002), loss of the antiandrogenic actions of progestins and estrogens, which are at low levels perimenstrually.

It is androgens, secreted by the ovaries and by the adrenal cortex, that seem to stimulate sexual desire in women. The androgens may exert this action by a combination of increased genital skin sensitivity (Salmon & Geist, 1943), a direct action on neurons in the spinal cord and brain, enzymatic conversion of androgens to estrogens (by a process termed "aromatization"), and the release of or changed responsiveness to other neurotransmitters (e.g., dopamine, norepinephrine, serotonin), which stimulate or inhibit the activity of other neurons, and neuromodulators (e.g., vasoactive intestinal peptide, endorphins), which change the responsiveness of neurons to neurotransmitters.

Androgens plus estrogens plus aromatization-produced estrogens have been shown to affect libido. In the laboratory rat, estrogen sensitizes the genital skin to touch (Komisaruk, Adler & Hutchison, 1972; Kow & Pfaff, 1973–74), but the precise mechanism for these effects is not known. Androgens also sustain the clitoris, labia, periurethral glands, nipples, and pelvic musculature (Levin, 2002).

The level of subjective sexual arousal evoked by erotic fantasy, film, and audiotape remains stable across phases of the menstrual cycle. This was demonstrated experimentally not only at the group level, in which mean ratings of subjective sexual arousal were compared across phases of the cycle, but at the individual level, through correlation among ratings by the same person at different cycle phases. Also, the relative ordering of women in terms of level of subjective sexual arousal remained stable across the phases of the menstrual cycle (Meuwissen & Over, 1992).

Thus, the menstrual cycle prepares the organs of the reproductive system for potential pregnancy in women's repeating monthly cycle over their reproductive years. By contrast with nonhuman mammalian species, in which female sexual behavior is more closely tied to the timing of ovulation, the significant cognitive complexity afforded by the human brain has created a more flexible and consequently more variable link between reproductive physiology and sexual behavior.

4

Are Orgasms Good for Your Health?

From the time of Hippocrates to Freud, doctors stimulated orgasms in women and girls as a clinical procedure known as "medical massage." Ostensibly performed to treat "hysteria," the objective of medical massage was to induce a "hysterical paroxysm," manifested by rapid respiration and pulse, reddening of the skin, vaginal lubrication, and abdominal contractions. In ancient Egypt and Greece, hysteria was believed to be the revolt of the uterus against sexual deprivation (*hyster* means "womb" in Greek). Rachel Maines (1989) described the late-nineteenth-century medical treatment of hysteria with electric vibrators. She reported that the electromechanical vibrator, introduced as a medical appliance in the 1880s, was an innovation designed to improve the efficiency of medical massage.

Medical massage disappeared from doctors' offices in

the 1930s. This was probably because hysteria was now treated by a novel procedure: psychotherapy. And it seems that when the use of mechanical stimulation became associated in the popular culture with sexual arousal and pleasure, the embarrassed medical establishment turned away from its use as a therapy (Blank, 1994).

Sexual health is now discussed more openly, after generations of being taboo. Not only is there interest in this subject in the medical and research communities, but the general public and the media are more openly discussing sexuality and sexual health. With the approval in 1998 of the use of sildenafil (Viagra) to treat male erectile dysfunction, research into male sexuality has increased. However, there is far less interest in, and far less literature on, female sexuality than male sexuality. An electronic search of the holdings of the National Library of Medicine yielded approximately 14,000 publications on male sexual disorders, but only 5,000 on female sexual disorders. A search using the words *sexual dysfunction* turned up 17,000 publications for men and 9,000 for women. Some claim that the assessment of sexual problems in women has been neglected due to the lack of sensitive and reliable outcome measures, because there is no defining, objective physical event with which to measure arousal or orgasm in women, as there is with penile erection in men (Rosen, 2002). We believe that in addition to the easier availability of objective measures of sexual response in men, this sex discrepancy is probably due to a traditional predominance of men as researchers in the field of sexual health and in the pharmaceutical industry.

Epidemiological data from the National Health and Social Life Survey (NHSLS) suggest that sexual problems affect 43 percent of women and 31 percent of men in the United States (Laumann, Paik & Rosen, 1999). These figures are based on a reanalysis of the data from 1,749 women and 1,410 men, aged 18 to 59, who responded to the 1992 NHSLS. Women who reported any sexual difficulty—such as lack of desire, difficulty in becoming aroused, inability to experience orgasm, anxiety about sexual performance, reaching orgasm too rapidly, pain during intercourse, or failure to derive pleasure from sex—were considered to have a sexual disorder. The researchers did not ask subjects about their level of distress, which is now believed to be a key component in the diagnosis of sexual disorder (K. P. Jones, Kingsberg & Whipple, 2005).

The International Consensus Development Conference on female sexual dysfunction, held on October 22, 1998 (reported by Basson, Berman, et al., 2000), added "personal distress" to the existing definitions of sexual disorders. On the basis of the findings of this panel, it became clear to other researchers and clinicians that women do not respond in the male-typical "linear," sequential pattern of desire, arousal, and orgasm, and that women's sexual response is more complex than men's. In addition, psychosocial factors may play a greater role in women's than in men's sexual response. Although more research is needed in this area, there is currently a drive to address women's sexual response more holistically than men's and more comprehensively than in the past.

From 1979 to 1983, Davey Smith, Frankel, and Yarnell (1997) conducted a study on the relationship between frequency of orgasm and mortality in the United Kingdom. At the ten-year follow-up, they found that the mortality risk was 50 percent lower among men who had frequent orgasms, defined in the study as two or more per week, than among men who had orgasms less than once a month. Even when controlling for other factors such as age, social class, and smoking status, a strong and statistically significant inverse relationship was found between orgasm frequency and risk of death—that is, the higher the frequency of orgasm, the lower the risk of death. The authors concluded that sexual activity seems to have a protective effect on men's health.

A research study on middle-aged men suggested a relationship between the levels of the hormone dehydroepiandrosterone (DHEA), which is released into the bloodstream at orgasm, and a reduction in the risk of heart disease (Feldman, Johannes, et al., 1998).

A study conducted from 1972 to 1975 examined the sex lives of a hundred Israeli women hospitalized with myocardial infarction (heart muscle destruction due to sudden blockage of a blood vessel that supplies the muscle), compared with a control group of a hundred women hospitalized for other reasons (Abramov, 1976). The groups were matched for age. Patients underwent a fifty-seven-item interview about their sex lives, including questions on the incidence of "frigidity" and the onset of menopause. ("Frigidity" in this survey included a lack of enjoyment of sexual intercourse, an inability to experience orgasm during intercourse that led to emotional distress, or a lack of sexual intercourse due to a partner's illness or erectile dysfunction.) The data showed a statistically significant positive correlation between frigidity and sexual dissatisfaction and between frigidity and a history of heart attack.

It has been suggested that sexual expression may lead to a decreased risk of cancer, because of the increased levels of oxytocin and DHEA associated with arousal and orgasm in women and men. Murrell (1995) suggested a preventative role for oxytocin in the development of breast cancer, in part by its action in aiding the elimination from the breast of fluid containing carcinogenic free radicals. And a study of male breast cancer conducted with 23 men in Greece found an inverse correlation between frequency of orgasm during adulthood and the incidence of this cancer (Petridou et al., 2000). The 23 men with breast cancer seemed to have had, on average, fewer orgasms than the 76 healthy but otherwise similar men in the comparison group. The authors speculated that reduced conversion of testosterone to the androgenic hormone dihydrotestosterone may play a role in increasing the likelihood of breast cancer in men.

On a less life-threatening level, there is evidence that orgasm may help people go to sleep. Orgasm causes a surge in oxytocin and endorphins that may have a sedative effect (Odent, 1999). One study found that 32 percent of 1,866 American women who reported masturbating in the previous three months did so to help them fall asleep (Ellison, 2000).

The relationship between orgasm and reproductive health was assessed in a retrospective case-control study of 2,012 women in the United States (Meaddough et al., 2002). The study examined the relationship between sexual response, orgasm, and the incidence of the often-painful disease endometriosis. The data showed that women who engaged, sometimes or often, in sexual activity during menstruation were less likely to develop endometriosis. These women were also more likely to report having orgasms during menstruation than those who rarely had intercourse when menstruating. The researchers concluded that sexual activity and orgasm during menstruation have a potentially protective effect against endometriosis.

Other researchers interviewed 1,853 pregnant women about their sex practices, including frequency of intercourse and experience of orgasm (Sayle et al., 2001). The women were at approximately twenty-eight weeks of gestation. Follow-up interviews were conducted later in the pregnancy and after delivery. Women who reported sexual intercourse with orgasm, sexual intercourse without orgasm, and orgasm without sexual intercourse were more likely to carry their pregnancy to full term than women who reported not engaging in any sexual activity late in their pregnancy. The authors suggested that continued sexual activity late in pregnancy may provide some protection against preterm delivery.

Ellison (2000) reported that orgasm may provide relief from pain. In the study, 9 percent of about 1,900 U.S. women who reported masturbating in the previous three months cited relief of menstrual cramps as their motivation. In a study of 83 women who suffered migraine, Evans and Couch (2001) reported that orgasm resulted in at least some relief for more than 50 percent of the women. Although relief of migraine through orgasm is less reliable and less effective than relief through drug therapies, the analgesic effect of orgasm is more rapid. Other research found that vaginal self-stimulation resulted in a significant increase in experimental pain thresholds (Komisaruk & Whipple, 1984; Whipple

& Komisaruk, 1985). Pain threshold was measured as the force at which a gradually increasing compression of the fingers became painful. Continuously maintained pressure stimulation on the anterior vaginal wall produced a more than 50 percent increase in the pain threshold (i.e., it produced analgesia). Comparable pressure stimulation of the posterior vaginal wall had no significant analgesic effect. In a second study, when the participants were asked to apply the vaginal self-stimulation in a way that felt pleasurable (rather than simply applying continuous pressure), significant analgesia, again measured as pain thresholds for finger compression, was elicited from stimulation of all the genital regions (Whipple & Komisaruk, 1988). In both studies, the greatest (more than 100%) increase in pain threshold occurred when the self-stimulation resulted in orgasm. The increased pain threshold was not associated with a change in touch thresholds, measured as sensitivity to gentle touch stimulation on the back of the hand. These findings indicate that vaginal self-stimulation produces analgesia rather than anesthesia (i.e., it suppresses pain without affecting sensitivity to touch).

Sexual activity and orgasm have been shown to reduce stress (Charnetski & Brennan, 2001). This may be due to the surge in oxytocin that accompanies sexual activity and, to a greater extent, orgasm. Increased levels of oxytocin are known to be associated with reduced stress and an altered response to stress; orgasm relieves tension as oxytocin stimulates feelings of warmth and relaxation (Weeks, 2002). In another study of 2,632 women in the United States, 39 percent of those who masturbated said they did so to relax (Ellison, 2000).

In men, two studies have provided evidence that a higher frequency of ejaculations over the years is correlated with a lower incidence of prostate cancer. In an Australian interview study of more than 2,000 under the age of 70 (Giles et al., 2003), those who recalled having had an average of four or more ejaculations per week during their twenties, thirties, and forties had a significantly lower (by one-third) risk of developing prostate cancer than men who reported an average of fewer than three ejaculations per week during the same age period. There was no association of prostate cancer with the number of sexual partners, suggesting that infectious factors did not account for the difference. In their analysis, the authors combined counts of ejaculations occurring during sexual intercourse and masturbation. While the authors raised the possibility of the mediation of these effects by hormonal factors, no correlations have been established between incidence of prostate cancer and any specific hormone. The mechanism for the effects observed in Giles et al.'s study remains unknown.

In a U.S. study (Leitzmann et al., 2004), relatively high ejaculation frequency (at least 21 per month compared with 7 or fewer per month) was related to a decreased risk of total prostate cancer (organ-contained and advanced cancer combined) in a questionnaire study of more than 50,000 men aged 40 to 75 years. Ejaculation frequency from sexual intercourse and masturbation was combined. In addition to considering possible mediating mechanisms of hormone levels and infection, the authors also speculated that ejaculations may clear the prostate of potential carcinogenic substances and that psychological stress reduction resulting from ejaculation could reduce the release of prostate cell–stimulating factors from nerves that supply the prostate.

There are several scientific review articles on the incidence of death during coitus, with a focus on men but little information on women. According to one medical examiner, "death in the saddle" follows a pattern in which "the deceased is usually married; he is with a nonspouse in unfamiliar surroundings after a big meal with alcohol" (A. W. Green, 1975). Garner and Allen (1989), who refer to the phenomenon also as "la mort d'amour," cite one study in which 14 of 20 cases of coital death occurred during extramarital intercourse, although those cases comprised only 0.3 percent of 5,559 cases of sudden death. They cite another study in which coroners estimated that "acute coronary insufficiency resulting from coitus is a fact, but the incidence rate is no more than three out of every 500 subjects with heart disease." Garner and Allen conclude that "it is obvious that coital death is a rare occurrence, and that reports of coital death of a middle aged, middle-class, male patient with heart disease who engages in sexual activity with his wife of 20 or more years in their own bedroom is even rarer."

In a more recent review (Safi & Stein, 2001), the authors caution that the increases in heart rate and blood pressure that occur during coitus might "precipitate the fracture or erosion of a vulnerable atherosclerotic plaque with subsequent thrombus [clot] formation and arterial occlusion." However, they cite a survey of more than 1,700 patients in which "sexual activity was noted to be a potential triggering event prior to MI [myocardial infarction] in only 1.5 percent of patients." These authors concluded that "absolute risk caused by sexual activity is considerably low: one chance in one million healthy individuals. Cardiac rehabilitation and exercise training programs can reduce incidence of angina pectoris during sexual activity."

These various studies, taken together, provide a fairly convincing answer to the question posed by this chapter: orgasms *are* good for your health, except in a small percentage of men (and perhaps women?) with heart disease.

5

When Things Go Wrong

A World Health Organization task force meeting in Geneva in 2002 developed some working definitions related to sex, sexuality, and sexual health that provide a useful context in which to discuss sex-related problems, including what can go wrong with orgasm. (All definitions quoted below are from www.who.int/reproductive-health/gender/sexual-health.html.)

Sex, the task force participants wrote, "refers to the biological characteristics, which define humans as female or male. (These sets of biological characteristics are not mutually exclusive as there are individuals who possess both, but these characteristics tend to differentiate humans as males and females. In general use in many languages, the term sex is often used to mean 'sexual activity,' but for technical purposes in the context of sexuality and sexual health discussions, the above definition is preferred.)"

Sexuality is "a central aspect of being human throughout life and encompasses sex, gender identities and roles, sexual orientation, eroticism, pleasure, intimacy and reproduction. Sexuality is experienced and expressed in thoughts, fantasies, desires, beliefs, attitudes, values, behaviors, practices, roles and relationships. While sexuality can include all of these dimensions, not all of them are always experienced or expressed. Sexuality is influenced by the interaction of biological, psychological, social, economic, political, cultural, ethical, legal, historical and religious and spiritual factors."

Sexual health is "a state of physical, emotional, mental and social well-being related to sexuality; it is not merely the absence of disease, dysfunction or infirmity. Sexual health requires a positive and respectful approach to sexuality and sexual relationships, as well as the possibility of having pleasurable and safe sexual experiences, free of coercion, discrimination and violence. For sexual health to be attained and maintained, the sexual rights of all persons must be respected, protected and fulfilled."

Sexual *dissatisfactions* and *disorders* do not necessarily imply that a man or woman is "dysfunctional"; the term *dysfunction* refers to a medical condition—one that can be treated medically. The sexual dysfunctions and disorders of men and women are currently defined in terms of three phases of the "sexual response cycle": desire, arousal, and orgasm, plus the sexual pain disorders. Most women do not fit into this linear model, as there is not a single sexual response cycle for women. However, there are commonalities between men and women. Their most common shared sexual complaints are as follows: (1) loss of sexual desire; (2) inadequate mental or genital arousal (resulting in erectile deficit in men and vaginal dryness in women); (3) impaired orgasm (in men, ejaculation may be premature, delayed, painful, or not experienced; in women, orgasm may be diminished, delayed, or not experienced); and (4) sexual pain disorders (more frequently reported by women, but occasionally reported by men) (Plaut, Graziottin & Heaton, 2004).

Hypersexuality, while quite rare, does occur in both men and women. The predominant complaint may be excessive desire or mental arousal, "sex addiction" or sexual compulsion, or unwanted persistent genital arousal without desire or mental arousal. Persistent sexual arousal syndrome was recently described in women (Plaut, Graziottin & Heaton, 2004).

The information necessary for evaluating and treating sexual disorders or dysfunctions includes: sexual history, medical history, psychosocial history, a focused physical examination, and recommended laboratory tests. Specialist referrals may be considered whenever a health care provider and patient believe this is appropriate. Individuals and their partners need to participate actively in decision making about treatment options (for further details, see Hatzichristou et al., 2004; Plaut, Graziottin & Heaton, 2004).

In considering sexual problems, one should keep in mind that these problems can affect both men and women, at any age, regardless of whether the person prefers partners of the same or the other sex and regardless of whether the person is otherwise healthy or has a chronic illness or disability (Whipple & Brash-McGreer, 1997).

Problems for Men

Sexuality is a complex biopsychosocial process. In men, the physiological aspects of sexual response, such as erection and ejaculation, need to be understood in the context of interpersonal, intrapersonal, and cultural factors. Problems with sexual function can be lifelong or acquired, global or situational. The etiology (cause and origin) may be organic, psychological, mixed, or unknown. Consideration must also be given to the individual's and his partner's needs and priorities, which will be influenced by cultural, social, ethnic, religious, and national factors. Men with sexual problems and their partners should select the best treatment for their sexual concerns, after receiving appropriate education that includes information on sexuality and all treatment options for the couple's particular concerns. Both the individual and his partner should always be involved in any therapeutic process (Lue et al., 2004). The definitions of male sexual dysfunctions used here derive from the report of the Second International Consultation on Erectile and Sexual Dysfunctions (later renamed the International Consultation on Sexual Medicine), held in June 2003, as reported by Lue et al. (2004).

It is important to be aware that many of the disorders mentioned in this chapter frequently coexist with other disorders and cannot be treated in isolation. Also, one must consider the various treatments for male sexual dysfunctions in the context of traditions, ethnicity, and socioeconomic conditions, as well as the individual's and his partner's preferences, expectations, and psychological status (Lue et al., 2004).

Erectile Dysfunction

Fugl-Meyer et al. (2004) summarized twenty-four studies on the prevalence of erectile dysfunction (ED), conducted between 1993 and 2003 in various countries. The prevalence in men below the age of 40 years was 1 to 9 percent; for men aged 40 to 59 years, it ranged from 2 to 9 percent to as high as 20 to 30 percent. Most of the studies found a rate of 20 to 40 percent for men aged 60 to 69 years. The prevalence rates for men in their seventies and eighties ranged from 50 to 75 percent (Lewis et al., 2004).

According to Lue et al. (2004), ED "is defined as the consistent or recurrent inability of a man to attain and/or maintain a penile erection sufficient for sexual activity." Erectile difficulties must occur consistently or recurrently to qualify for a diagnosis of ED, with a three-month minimum duration generally meeting the accepted criteria for diagnosis. In some instances of trauma or surgically induced ED (e.g., after a radical prostatectomy), the diagnosis may be given before three months.

A diagnosis of ED should be made by a health care professional who uses a multidisciplinary approach. All patients reporting presumptive ED should undergo a medical, sexual, and psychosocial history, physical examination, and laboratory tests (Lue et al., 2004). The health professional should evaluate factors such as endocrine problems, circulatory problems, diabetes, cardiovascular problems, obesity, cigarette smoking, alcoholism, substance abuse, relationship problems, depression, and other psychosexual problems.

The use of prescription and nonprescription drugs such as antihypertensive agents, psychotropic drugs, antiarrhythmics, antiandrogens, and steroids may contribute to ED. Whenever medical therapy is indicated, prior or concurrent psychosexual counseling is advisable, such as by an AASECT-certified sex therapist (AASECT is the American Association of Sexuality Educators, Counselors, and Therapists; for a list of certified sex therapists, see www.aasect.org). Once the cause of ED is determined, there are multiple treatment options; discussion of these is beyond the scope of this book. Simply filling a prescription for sildenafil does not generally constitute proper evaluation and treatment of ED.

Premature Ejaculation

The prevalence rates for early, rapid, or premature ejaculation (PE), based on five descriptive studies, range from 9 to 31 percent (Lewis et al., 2004). According to Lue et al. (2004), "Premature ejaculation, also referred to as a rapid or early ejaculation, is defined according to three essential criteria: (a) brief ejaculatory latency, (b) loss of control, and (c) psychological distress in the patient and/or partner." Ejaculatory latency, the time from the onset of intercourse to ejaculation, of two minutes or less may lead to a diagnosis of PE; the diagnosis should also include consistent inability to delay or control ejaculation and marked distress about the condition. Subtypes of PE are lifelong versus acquired, global versus situational, and the co-occurrence of other sexual problems, particularly ED. Lue et al. (2004) state that "about 30 percent of men with PE have co-occurring ED, which typically results in early ejaculation without full erections." PE may be associated with sexual problems in the female partner, particularly anorgasmia (inability to experience orgasm) or sexual pain disorder. In such cases, the couple may find it beneficial to consider sex therapy.

As with ED, a man experiencing PE should undergo a medical and sexual history, physical examination, and investigation of anxiety and interpersonal factors. Men with lifelong PE are now being treated pharmacologically with "off-label" selective serotonin reuptake inhibitors (SSRIs) and topical local anesthetics, as well as phosphodiesterase-5 (PDE-5) inhibitors (sildenafil-type drugs). (The term *off-label* is used when a drug is given to a type of patient or for a reason other than those approved by the U.S. Food and Drug Administration [FDA].) Men with acquired or situational PE are treated with pharmacotherapy or behavioral therapy or both, according to the individual's and his partner's preference (Lue et al., 2004). Behavioral therapy usually includes the "stop-start" or the squeeze techniques. Both techniques are based on the idea that PE occurs because the man fails to pay sufficient attention to pre-orgasmic levels of sexual tension (Semans, 1956; Masters & Johnson, 1970). However, there are as yet no long-term outcome data for evaluating these treatment methods (De Amicis et al., 1985).

Male Orgasmic Dysfunction

Both organic and psychogenic factors can contribute to male orgasmic dysfunction (MOD). According to Lue et al. (2004), MOD includes "a spectrum of disorders ranging from delayed ejaculation to a complete inability to ejaculate, anejaculation, and retrograde ejaculation . . . Any medical disease, drug or surgical procedure that interferes with either central control of ejaculation or the peripheral sympathetic nerve supply to the vas [deferens] and bladder neck, the somatic efferent [output to voluntary muscle] nerve supply to the pelvic floor or the somatic afferent [sensory input from the skin] nerve supply to the penis can result in delayed ejaculation, anejaculation, and anorgasmia."

Four neurophysiological tests are routinely used to assess men with MOD: pudendal somatosensory evoked potentials, motor evoked potentials, sacral reflex arc testing, and sympathetic skin responses—each of which tests the integrity of different specific links in the chain of processes by which penile stimulation elicits erection, orgasm, and ejaculation (McMahon et al., 2004). Men who never experience orgasm and ejaculation may suffer from a biogenic failure of emission or psychogenic-inhibited ejaculation. Men who only occasionally experience orgasm and ejaculation may have psychogenic-inhibited ejaculation or penile hypesthesia (subnormal sensitivity), which is associated with age-related degeneration of the afferent (sensory) penile nerves (McMahon et al., 2004). Therapy involves treating any concomitant medical problem, as well as behavioral therapy.

One type of MOD, inhibited ejaculation, becomes more prevalent (as do other sexual dysfunctions) as men age (Feldman, Goldstein, et al., 1994). Inhibited ejaculation may be associated with cultural and religious beliefs, concurrent psychopathology such as unconscious aggression and unexpressed anger, insufficient sexual arousal, a preference for masturbation over partnered sex, fear of inducing pregnancy, or fear of sexually transmitted infection (Perelman, 2001).

Priapism

As noted by Lue et al. (2004), "Priapism is a relatively rare condition in men. It is defined as an unwanted erection, not associated with sexual desire or sexual stimulation, and lasting for more than 4 hours." There are three different types of priapism, although there may be some overlap among them. Low-flow, or ischemic (insufficient blood supply), priapism is the most common form and is associated with a failure of detumescence, anoxia (lack of blood-borne oxygen supply to the tissue), and ultimately, if untreated, necrosis (cell death) of the muscle of the corpora cavernosa. High-flow, or well-oxygenated, priapism is less common and may occur after surgical treatment or pelvic trauma. The third type is recurrent, or stuttering, priapism, which commonly occurs in men with sickle-cell disease. Recurrent priapism is usually high flow, but may become low flow and anoxic (Lue et al., 2004).

Cold showers or ice packs may be beneficial during the early stages of priapism. Exercise and urination are occasionally helpful. If ischemic priapism is diagnosed, it is essential to decompress the corpora cavernosa as soon as possible by withdrawing blood from the penis with a hypodermic needle. Pharmacological agents can be used to treat the other types of priapism.

Peyronie's Disease

Peyronie's disease is an acquired disorder of the tunica albuginea, the tough connective tissue capsule enclosing the corpora cavernosa that enables the penis to become rigid when the corpora cavernosa are engorged with blood. The disease is characterized by the formation of a plaque of fibrous tissue in the tunica albuginea and is often accompanied by penile pain and deformity on erection. There may be difficulty with penetration as a result of the curvature, and the condition may be accompanied by some impairment of erectile capacity (Lue et al., 2004). Part of the medical history taken by the health care professional should include information on the individual's ability to have and maintain an erection. Most men require no more than education and reassurance.

Desire Disorder

The report of the Second International Consultation on Erectile and Sexual Dysfunctions did not include the category of sexual desire disorders in men. An open question is whether these disorders should be classified as a sexual *dysfunction*, as they are in the *Diagnostic and Statistical Manual of Mental Disorders* (DSM-IV;

American Psychiatric Association, 1994), the diagnostic "bible" of psychiatrists and psychologists (Maurice, 1999). One study reported that 33 percent of women and 16 percent of men claimed that during the prior twelve months, there had been a period of several months or more when they lacked interest in having sex (Laumann et al., 1994). The results of other studies support this finding (e.g., Frank, Anderson & Rubenstein, 1978).

Desire disorders fall into two categories: hypoactive sexual desire disorder (HSDD) and sexual aversion disorder (SAD). According to the DSM-IV, the criteria for a diagnosis of HSDD are "persistent or recurrently deficient (or absent) sexual fantasies and desire for sexual activity that causes marked distress or interpersonal difficulty." The criteria for SAD are "(a) persistent or extreme aversion to, and avoidance of, any genital contact with a sexual partner, causing marked distress or interpersonal difficulty, and [that] (b) the disturbance does not occur exclusively during the course of another mental disorder."

Here we focus on HSDD, which is often not clinically distinguished from other problems of sexual function. One situation in which a sexual desire problem becomes evident is when two partners are sexually interested but not at the same level, a condition termed sexual desire discrepancy. Again, is this really a sexual dysfunction or a sexual disorder? Sometimes a desire discrepancy does become a problem for which a couple or an individual consults a health professional. The health professional must then determine whether the problem is acquired or lifelong, situational or generalized.

Male disorders of desire tend to increase with age. The most well-established etiologies are depression/anxiety, hypogonadism (including andropause, or "male menopause"), and hyperprolactinemia (an excessive secretion of the hormone prolactin) (Plaut, Graziottin & Heaton, 2004). Most of the evidence relating a decrease in sexual desire to hormone levels involves decreased androgen levels resulting from surgical or chemical castration, aging, or hypogonadal states (Maurice, 1999). Nonhormonal factors that may affect desire in men are marital boredom, diminished partner attractiveness, and chronic illness.

Many commonly used pharmacological agents may decrease sexual desire, such as antiandrogenic drugs, psychoactive drugs, antihypertensive agents, cardiac drugs (including beta-adrenergic blockers and calcium blockers), drugs that bind with testosterone, cimetidine (taken for the treatment of peptic ulcers), dichlorphenamide and methazolamide (both drugs for treating glaucoma), cholesterol-lowering drugs, and steroids such as prednisone and dexamethasone (Maurice, 1999). An individual's medical and medication history must be considered before any treatment for HSDD is prescribed.

Individuals may also have low desire as a result of other sexual disorders, such as ED. Testosterone should not be used routinely in an attempt to increase sexual desire in the absence of indications of low testosterone levels. Testosterone treatment is associated with potential risks, especially the possibility of exacerbating unrecognized prostate cancer. Multiple factors are usually involved in male desire disorder, and these men should be treated with a multidisciplinary approach.

Challenging Beliefs about Female Sexuality

Participants at the International Consultation on Sexual Medicine, held in Paris in June 2003, challenged some of the common beliefs about women's sexual response (Basson, Leiblum, et al., 2003). Here we briefly summarize these beliefs and the challenges to them.

—Belief 1: *Organic sexual problems can be separated from psychogenic problems.* Sexual disorders in women may involve multiple psychological, interpersonal, and biological/organic causes, and these influences are not always separate entities.

—Belief 2: *The primary reason women engage in sexual be-*

havior is conscious or subliminal awareness of sexual desire (e.g., sexual thoughts or sexual fantasies). Women seem to be motivated to have sex for highly complex and varied reasons. Women in new relationships are more likely to experience spontaneous desire—in the form of sexual thoughts and fantasies—than are women in established relationships, who may think of sex infrequently.

—Belief 3: *Sexual desire always precedes sexual arousal.* Arousal often occurs *before* desire for women, or women may experience desire and arousal simultaneously. Again, desire is not the only reason that women engage in sexual activity; they have a wide variety of other motives, including a wish to be intimate with their partner. According to Basson, Althof, et al. (2004), "Desire is consequently experienced after arousal such that continued arousal and a responsive type of desire coexist and reinforce each other in keeping with the conceptualization of women's sexual response."

—Belief 4: *Women's sexual arousal can be characterized by genital vasocongestion, vaginal lubrication, and an awareness of genital throbbing and tingling.* Many women who have genital signs of arousal do not feel subjectively aroused, and many women may not even be aware of the physiological changes in their bodies when they are aroused. Even if they are aware of genital and breast vasocongestion, the changes may not correlate with increased vaginal engorgement. (Engorgement can be measured with a vaginal photoplethysmograph, a vaginal probe with a light source and light detector that measures changes in the amount of light reflected through the vaginal wall as an indication of changes in vaginal blood flow.) Still, vaginal lubrication commonly occurs during sexual stimulation even when a woman does not desire or enjoy the stimulation.

—Belief 5: *The sexual response of women remains stable over time and circumstance.* The sexual response of women varies naturally over the lifespan and is influenced by a host of factors, including the context of sexual interactions, pregnancy and menopause, medical conditions, and psychological factors (most notably the interpersonal relationship). Research suggests that a normative, gradual decline in sexual interest and response occurs with aging and natural menopause.

—Belief 6: *Women feel distress when they experience changes in their sexual response.* Many women do not feel distress when they lose interest in sex or experience a lack of response. And unless women do feel distress, these are not really problems and are of little clinical relevance (Basson, Leiblum, et al., 2003).

Problems for Women

Research on women's sexual problems and concerns is lagging behind research on men's sexual problems by a 2:1 or 3:1 ratio. Researchers and health care providers are now beginning to recognize that sexual response in men and women differs in significant ways. The diagnostic categories of women's sexual "dysfunction" as presented in the DSM-IV are based on a conceptualization of sexual response depicted by Masters and Johnson (1966, 1970) and Kaplan (1974). For their time, their work was a great advance in the knowledge of human sexuality. However, the concept of one linear sequence of predominantly genitally focused events has not proven helpful in assessing and managing women's sexual problems and disorders (Basson, 2001).

Several models have been proposed, each of which views women's sexual response as more of a circular than a linear pattern (Whipple & Brash-McGreer, 1997; Basson, 2001; Plaut, Grazziottin & Heaton, 2004). A full appreciation of women's sexual function requires an understanding of the underlying psychosocial, as well as physiological, components. When assessing women's concerns, the health care provider's focus must extend beyond physical

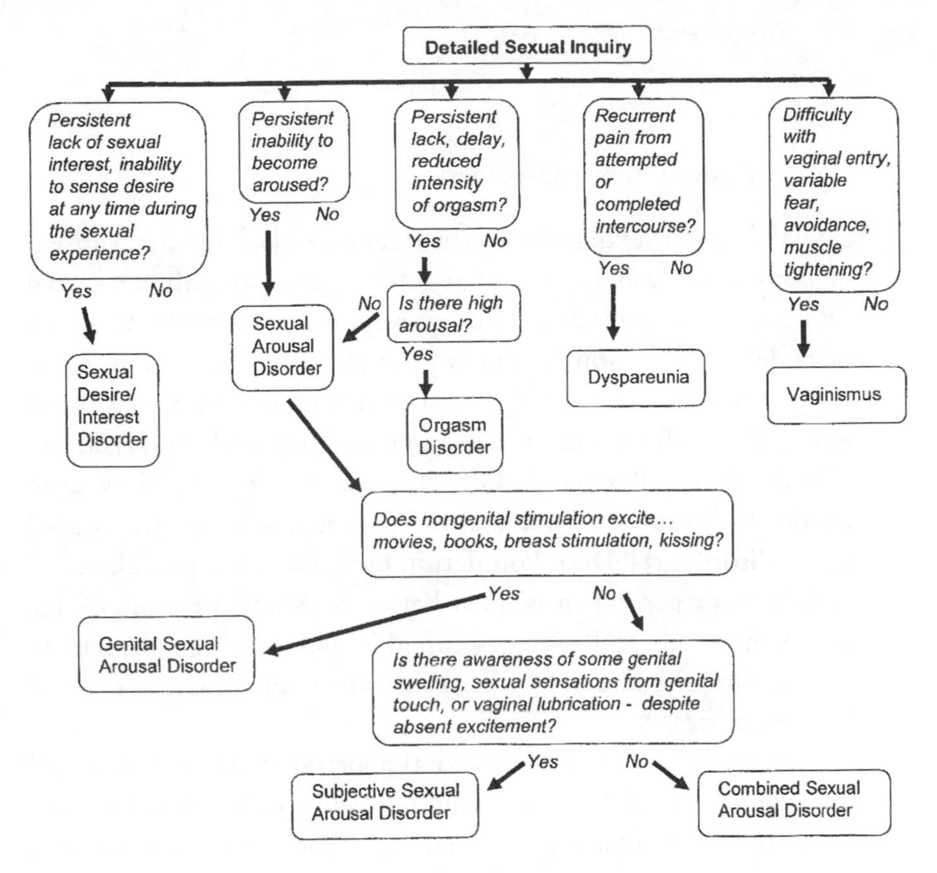

Guide for characterizing a sexual disorder. (Adapted with permission from R. Basson, W. C. M. Weijmar Schultz, et al., 2004)

issues to encompass the emotional and relationship milieu in which the concerns exist (K. P. Jones, Kingsberg & Whipple, 2005). As with men, problems with women's sexual function can be lifelong or acquired, global or situational. The etiology may be organic, psychogenic, mixed, or unknown.

We use here the new categories of sexual disorders in women proposed by the International Consultation in June 2003; these categories have not yet been incorporated into the DSM or the *International Statistical Classification of Disease and Related Health Problems* (ICD-10; World Health Organization, 1992).

Sexual Desire/Interest Disorder

Sexual desire or sexual interest disorder is newly defined as "absent or diminished feelings of sexual interest or desire, absent sexual thoughts or fantasies and a lack of responsive desire. Motivations (here defined as reasons/incentives) for attempting to become sexually aroused are scarce or absent. The lack of interest is considered to be beyond the normative lessening with life-cycle and relationship duration" (Basson, Althof, et al., 2004). The DSM-IV term for desire disorders, as noted above for men, is hypoactive sexual desire disorder (HSDD). Population data indicate a prevalence of HSDD of 33 percent in women between 18 and 59 years of age (Laumann, Paik & Rosen, 1999); this may reach 45 percent in clinical samples, especially after menopause (Plaut, Grazziottin & Heaton, 2004).

The term *libido* refers to a sexual appetite, desire, drive, sexual impulse, and sexual interest that motivates individuals to focus their attention on having a sexual experience. Plaut, Grazziottin, and Heaton (2004) suggest three basic components to desire: biological, motivational, and cognitive.

Biologically, desire is hormonally influenced. Hormones may affect the intensity of desire and sexual response. The primary hormonal etiologies of sexual desire disorder are hypoestrogenism, androgen insufficiency syndrome, hyperprolactinemia, and hypothyroidism. Alcohol addiction and smoking may contribute to sexual disorders, as may certain disease processes and medications (Plaut, Grazziottin & Heaton, 2004). Medications that may cause a loss of desire are SSRIs, antihypertensives, estrogen therapies, and corticosteroids. Also, a sudden drop in testosterone levels, as occurs with surgical menopause, can cause a lack of sexual desire (Whipple & Brash-McGreer, 1997; Kingsberg, 2002).

Motivationally, the need for intimacy, which seems to be particularly important to women, may contribute to and modulate sexual desire. The affective disorders of depression and anxiety seem to decrease sexual desire (Plaut, Grazziottin & Heaton, 2004). Cognitive factors may overlap with biological and motivational factors to diminish sexual desire. There are other reasons, beyond awareness of sexual desire, that motivate women to agree to or initiate sexual interactions with a partner. These are currently being investigated in a study of women of different ethnic backgrounds in North America. Some reasons already identified are: to express love, for pleasure, because the partner wants it, and to release tensions. Also identified are some reasons for *not* being sexually active, including no partner, no interest, too tired, partner has no interest, partner is too tired, own physical problem, and partner's physical problem (Cain et al., 2003). Basson, Althof, et al. (2004) report multiple reasons or incentives for a woman to be aroused and that once she experiences arousal, if it continues sufficiently long and is enjoyable, she may experience sexual desire.

Studies also indicate that sexually healthy women, particularly those in long-term relationships, are frequently unresponsive to spontaneous sexual thoughts (Bancroft, Loftus & Long, 2003). The Massachusetts Women's Health Study II suggested that most women experiencing decreased sexual drive are married, have symptoms of psychological pathology, smoke cigarettes, and are in perimenopause (Avis et al., 2000).

Cognitive factors affecting desire may be many. Indeed, the treatment of disorders of desire is usually cognitive-behavioral therapy, sex therapy, or psychodynamic treatment. Although cognitive-behavioral therapy is widely used, very few controlled trials of its alleged benefits have been conducted. There is some empirical support for the usefulness of sex therapy with sensate focus, but to our knowledge there are no studies on psychodynamic treatment (Basson, Althof, et al., 2004). However, because loss of desire is often related to interpersonal problems rather than biological factors, counseling for relationship and psychological issues should be considered before pharmacotherapy.

Currently, there are no drugs specifically prescribed for the treatment of any female sexual disorder, including disorders of desire. Several potential therapies are now in clinical trials. Estrogen therapies can help with physical problems related to menopause, but they have little effect in treating decreased desire (Suckling, Lethaby & Kennedy, 2003). Testosterone therapy has been used "off-label" for women with low sexual desire. Some women take oral estrogen and testosterone for low desire, but this treatment was approved only for menopausal symptoms. Other women take testosterone prepared by a compounding pharmacist, and others use a piece of the male testosterone patch or testosterone gel developed for men. None of these treatments is approved for women by the FDA. A recent study by S. R. Davis, Davison, et al. (2005) found no correlation between circulating androgen levels and sexual function in women: "no single androgen level is predictive of low female sexual function, and the majority of women with low dehydroepiandrosterone sulfate levels did not have low sexual function." These authors stated that their results are not in conflict with testosterone being used pharmacologically to treat HSDD, nor do their data provide support for efficacy of the therapy.

In December 2004, the Procter and Gamble company presented to the FDA its positive data on the effects of a testosterone patch (Intrinsa) developed for HSDD in surgically postmenopausal women (Shifren et al., 2000; Goldstat et al., 2003). However, the FDA Advisory Committee voted to send Intrinsa back for more studies, citing insufficient long-term safety data to support approval of the drug.

It is interesting that Viagra was approved for use in men in 1998 based on only six months of clinical trials, despite concerns about an increased risk of a fatal reaction if taken together with a nitroglycerin medication or with well-known recreational drugs (Kingsberg & Whipple, 2005). By contrast, after more than three years of study, Intrinsa was not approved for use in women, and long-term safety data are now requested. This difference suggests the existence of a male-female double standard in the treatment of sexual disorder.

Subjective Sexual Arousal Disorder

Subjective sexual arousal disorder is defined as the "absence of or markedly diminished feelings of sexual arousal (sexual excitement and sexual pleasure) from any type of sexual stimulation. Vaginal lubrication or other signs of physical response still occur" (Basson, Althof, et al., 2004). The International Consultation panel developed this new category based on data suggesting that most women who complain of arousal problems have genital vasocongestion comparable to that of women who do not complain of a loss of subjective arousal (K. P. Jones, Kingsberg & Whipple, 2005).

Genital Sexual Arousal Disorder

The International Consultation panel defined genital sexual arousal disorder as "complaints of impaired genital sexual arousal. Self-report may include minimal vulvar swelling or vaginal lubrication from any type of sexual stimulation and reduced sexual sensations from caressing genitalia. Subjective sexual excitement still occurs from non-genital sexual stimuli" (Basson, Althof, et al., 2004). A woman diagnosed with this disorder can still be subjectively aroused by sexual stimulation, but she has a marked loss of intensity of any genital response, including orgasm. Awareness of throbbing/swelling/lubrication is absent or markedly diminished. It is the woman's self-report of absent or impaired genital congestion and lubrication that is the basis of the definition of this disorder (Basson, Althof, et al., 2004). This diagnosis pertains mostly to women with autonomic nerve damage and estrogen deficiency who do not experience vasocongestion (K. P. Jones, Kingsberg & Whipple, 2005).

Combined Genital and Subjective Arousal Disorder

Combined genital and subjective arousal disorder is defined as "absent or markedly diminished feelings of sexual arousal (sexual excitement and sexual pleasure), from any type of sexual stimulation as well as complaints of absent or impaired genital sexual arousal (vulval swelling, lubrication)" (Basson, Althof, et al., 2004). The International Consultation panel noted that this is the most common clinical presentation for female arousal disorders. The woman also complains of a lack of sexual desire (K. P. Jones, Kingsberg & Whipple, 2005). The lack of subjective excitement from any type of sexual stimulation distinguishes these women from those with genital arousal disorder.

Persistent Genital Arousal Disorder

A provisional definition of persistent genital arousal disorder is "spontaneous, intrusive and unwanted genital arousal, e.g., tingling, throbbing, pulsating, in the absence of sexual interest and desire. Any awareness of subjective arousal is typically but not invariably unpleasant. The arousal is unrelieved by one or more orgasms and the feeling of arousal persists for hours or days" (Basson, Althof, et al., 2004). This provisional definition will allow further investigation of the prevalence and etiology of this poorly understood disorder. It may not be as rare as previously believed.

Sexual Aversion Disorder

Sexual aversion disorder is defined as "extreme anxiety and/or disgust at the anticipation of, or attempt to have, any sexual activity . . . Many clinicians feel the syndrome of extreme anxiety/panic associated with activation of the autonomic nervous system is a form of phobic reaction. However, the panel felt that the sexual context and sexual repercussions warrant its inclusion as a sexual disorder" (Basson, Althof, et al., 2004).

Orgasmic Disorder

Given that most women are unable to experience orgasm from intercourse alone and require extended clitoral or vaginal (G spot) stimulation (Whipple & Brash-McGreer, 1997; N. A. Phillips, 2000), we can assume there is a sizeable subset of women who have difficulty experiencing orgasm. Women's orgasmic disorder is defined as follows: "Despite the self-report of high sexual arousal/excitement, there is either lack of orgasm, markedly diminished intensity of orgasmic sensations or marked delay of orgasm from any kind of stimulation" (Basson, Althof, et al., 2004). Anorgasmia is a common problem that affects an estimated 24 to 37 percent of women (Rosen, 2000). In primary orgasmic disorder, a woman has never experienced orgasm through any means of stimulation; in secondary orgasmic disorder, a woman is anorgasmic after having once been orgasmic, or is orgasmic under some conditions but not others. Thus, secondary anorgasmia can be classified as generalized or situational, such as when a woman can experience orgasm by masturbation but not with a partner (K. P. Jones, Kingsberg & Whipple, 2005).

Some evidence was recently presented of a genetic factor playing a minor role in orgasmic response. Based on a questionnaire study of "sexual problems" in a sample of female twins in England, comparing identical and fraternal twins, the researchers concluded that there is a significant heritable component for "difficulty reaching orgasm during intercourse" (K. M. Dunn, Cherkas & Spector, 2005). The underlying mechanism is unknown.

The International Consultation panel developed the new definition of orgasmic disorder because the old definitions often ignored the criterion of "high" or "adequate" sexual arousal. The current definition incorporates the criterion that the woman has no problem becoming aroused. Anorgasmia is considered to be a disorder only if the woman finds it distressing.

Orgasm is more than simply a reflex. While it incorporates reflexive components, it also includes perception, which is not a necessary component of true reflexes. Orgasm may be triggered by a number of physical and mental stimuli. It does not even require direct genital stimulation. Mental (imagery-induced) orgasm in women has been demonstrated under laboratory conditions (Whipple, Ogden & Komisaruk, 1992; Komisaruk & Whipple, 2005).

Psychosexual issues are more frequently in play for women with lifelong orgasmic difficulties. Biologically, the coexistence of two or more disease processes and pharmacological side effects becomes increasingly likely with advancing age (Plaut, Graziottin & Heaton, 2004).

In the 1970s, the small-group format for the treatment of orgasmic disorders became popular, and books and videos were developed to "give women permission to experience orgasm" (a 1970s expression) and to share ways, in small "preorgasmic groups," to experience sensual and sexual pleasure (Barbach, 1975). Women were taught about directed masturbation for treating lifelong generalized orgasmic disorder. According to Meston, Hull, et al. (2004), "there are no consistent, empirical findings that psychosocial factors alone differentiate orgasmic from anorgasmic women." The most common methods used to treat anorgasmia are cognitive-behavioral, pharmacological, and systems theory approaches. Meston, Hull, et al. (2004) note that "cognitive behavioral therapy for anorgasmia focuses on promoting changes in attitudes and sexually-relevant thoughts, decreasing anxiety, and increasing orgasmic ability and satisfaction."

Behavioral exercises to deal with female orgasmic disorder include, as noted above, directed masturbation, with and without vibrators, which has been shown to be effective in groups and individually. If a woman is able to experience orgasm through masturbation but not with a partner (and experiencing orgasm with a partner is her desire), couple therapy may be helpful, after issues of anxiety, communication, trust, and past history have been addressed. Another behavioral approach often recommended is the Kegel pelvic floor muscle-strengthening exercises. Physical therapy and biofeedback methods are helpful in the correct use of these exercises (Ladas, Whipple & Perry, 2005). Graber and Kline-Graber (1979) found a positive correlation between the strength of a woman's pelvic muscles and the intensity of her orgasmic response. In their retrospective study (a study that asks participants to recall past events), these authors found that women with very weak pelvic muscles were anorgasmic.

Sensate focus exercises were developed by Masters and Johnson (1970) to reduce anxiety through the use of a series of body-touching exercises, moving from sensual-nonsexual (e.g., scalp and face massage) to increasingly sensual-sexual (e.g., genital massage). These exercises are still widely used today. However, according to Meston, Hull, et al. (2004), there are no reports that these exercises produce any substantial improvement in orgasmic response

A health care provider's "granting permission" and providing usable information to the woman with anorgasmia may be the most helpful treatment. And given that orgasm is not always essential to sexual satisfaction, and that the inability to experience orgasm during intercourse is not abnormal, one needs to consider whether or not an orgasm difficulty needs a closer look. If not experiencing an orgasm causes a woman distress, then it can be considered a sexual problem. It is important for women to know that they are in charge of their own orgasm; they can *have* an orgasm, but no one else can *give* them an orgasm. Women are responsible for their own pleasure and satisfaction. In a laboratory study, a woman who had a complete spinal cord injury chose to participate so as to learn more about her own sexual response (Whipple, Gerdes & Komisaruk, 1996). She had not experienced an orgasm during the two years since her injury. She had been told by a health care practitioner that since she had no movement or feeling below her breast, she could not be orgasmic again, so she had not tried any form of stimulation. During the three twelve-minute periods in the laboratory that included sexual self-stimulation (genital and nongenital), the woman experienced six orgasms.

Pain Disorders: Dyspareunia and Vaginismus

Dyspareunia is defined as "persistent or recurrent pain with attempted or complete vaginal entry and/or penile vaginal intercourse" (Basson, Althof, et al., 2004). The International Consultation panel updated the definition to include pain during penetration, not just during *attempts* at penetration. Dyspareunia is estimated to affect 14 percent of women annually, according to the National Health and Social Life Survey (Laumann, Paik & Rosen, 1999). The most common cause of sexual pain disorders among middle-aged and older women is atrophic vaginitis (vaginal inflammation related to wasting [atrophy] of the tissue). In a postmenopausal population in the Netherlands, 27 percent of the women surveyed reported vaginal dryness, soreness, and dyspareunia (Van Geelen, van de Weijer & Arnolds, 1996).

While psychological and relationship factors play an important role in dyspareunia, it is the only female sexual disorder in which organic factors contribute significantly (Anastasiadis et al., 2002). For treatment to be effective, it is important to determine at the outset whether the dyspareunia is lifelong or acquired. Introital dyspareunia (pain on insertion of the penis into the introitus, the entrance into the vagina) is usually caused by poor arousal, vestibulitis (inflammation of the vaginal vestibule), vulvar dystrophy (vulvar abnormality), perineal surgery, pudendal nerve entrapment, or pudendal neuralgia. Midvaginal pain is usually due to levator ani myalgia (pain in the pelvic floor muscle that elevates the anus). Deep vaginal pain may be caused by endometriosis, pelvic inflammatory disease, pelvic varicocele (varicose blood vessels), adhesions, referred abdominal pain, outcomes of radiotherapy, or abdominal cutaneous nerve entrapment syndrome (Plaut, Graziottin & Heaton, 2004).

The other type of pain disorder, vaginismus, is defined as "the persistent or recurrent difficulties of the woman to allow vaginal entry of a penis, a finger, and/or any other object, despite the woman's expressed wish to do so. There is often (phobic) avoidance, involuntary pelvic muscle contraction and anticipation/fear/experience of pain. Structural or other physical abnormalities must be ruled out/addressed" (Basson, Althof, et al., 2004). The International Consultation panel revised the definition because vaginal spasm, included in the earlier definitions, has never been documented. The panelists specified that involuntary contractions *may* occur. They noted that vaginismus typically prevents the full entry of a penis (or other object) but that vaginal entry can occur and causes discomfort and pain (K. P. Jones, Kingsberg & Whipple, 2005). Vaginismus affects 15 to 17 percent of women who consult sex therapy clinics (Spector & Carey, 1990; Anastasiadis et al., 2002).

There have been very few controlled studies of dyspareunia or vaginismus. A multidisciplinary approach for sexual pain is recommended for women with these disorders, with attention to the experience of pain, the emotional/psychological profile, any past genital mutilation or sexual abuse, and examination of the mucous membrane and the pelvic floor (Basson, Althof, et al., 2004). Examination of the genitals should be approached with gentleness and constant interaction with the woman about painful areas. It may be difficult or impossible to examine the vagina with a speculum.

Vaginal estrogen and vaginal lubricants can be used to enhance comfort with penetration for women with vaginal atrophy. Beyond treatment of medical conditions, such as atrophic vaginitis and endometriosis, women suffering from sexual pain disorders may benefit from psychological counseling and education. Instruction in progressive muscle relaxation and the use of vaginal dilators may be helpful. Although many clinicians define vaginal penetration as the goal of therapy, we suggest sexual pleasure for the woman and her partner is a better measure.

6

Diseases That Affect Orgasm

Many types of medical conditions can affect sexuality. These include: (1) neurological disorders such as head injury, multiple sclerosis, psychomotor epilepsy, spinal cord injury, and stroke; (2) vascular disorders such as hypertension and other cardiovascular diseases, leukemia, and sickle-cell disease; (3) endocrine disorders such as diabetes, hepatitis, and kidney disease; (4) debilitating diseases such as cancer, degenerative diseases, and lung disease; (5) psychiatric disorders such as anxiety and depression; and (6) voiding disorders such as overactive bladder and stress urinary incontinence (Whipple & Brash-McGreer, 1997; N. A. Phillips, 2000; K. P. Jones et al., 2005).

We focus here on some of the major diseases that more commonly result in problems related to orgasm. In general, any disease, injury, or surgery that affects the brain or spinal cord, or the nerves that serve them, can result in sexual impairment. Furthermore, many of the medications used to treat mental and physical conditions also affect sexual responsiveness. Since biological and psychosocial factors interact, there is often no clear identification of cause. It is important to remember that sexual problems can affect both men and women regardless of whether they have a chronic illness or disability, or no medical problem at all (Whipple & Brash-McGreer, 1997).

Diabetes Mellitus

Diabetes in men has long been recognized as accompanied by sexual problems. The Persian philosopher, physician, and scientist Avicenna (980–1037) was the first to mention a "collapse of sexual function" as a complication of diabetes. And by 1906, "impotence" was reported to be one of the most common symptoms of diabetes (Macfarlane et al., 1977). After the early 1900s, sexuality became a taboo topic in some western societies, and the relationship between diabetes and sexuality was hardly mentioned in the United States until after the Second World War. At that time, doctors advised that two people with diabetes should not marry and have children, and that the less attention given to impotence, the better (Enzlin, Mathieu & Demytteanere, 2003).

The sexual problems of women with diabetes were not mentioned until the discovery of insulin in 1921, when women who took insulin became healthy enough to become pregnant. In the years that followed, there have been many more studies on the sexual effects of diabetes in men than in women.

For men with diabetes, erectile dysfunction (ED) is a common problem; they often develop ED five to ten years earlier than men without diabetes (American Diabetes Association, 2001). The prevalence of erectile problems in men with diabetes has been reported to range from 27 to 71 percent (Schiavi, Stimmel, et al., 1993). Erectile problems may be a manifestation of microvascular (small blood vessel) or macrovascular (large blood vessel) disease (Maurice, 1999). A study of neurological factors in the etiology of ED was conducted in the Netherlands (Bemelmans et al., 1994). The study sample included three groups: 27 men with diabetes and ED, 30 men with diabetes but no ED, and 102 nondiabetic men with ED. The neurophysiological evaluations consisted of an assessment of somatic and autonomic nerves by measuring the latencies of somatosensory evoked potentials (a measure of the rate at which nerve impulses are transmitted) of the posterior tibial and pudendal nerves and the bulbocavernosus and urethra-anal reflexes. The researchers found a tendency toward a higher, but *not* statistically significant, incidence of sensory neuropathy (nerve pathology) in the men with diabetes and ED. In these men, pharmacological tests involving the corpora cavernosa suggested vasculogenic ED (i.e., ED of blood vessel origin). No endocrinological (hormonal) differences were found among the three groups in the study. The authors concluded that urogenital sensory neuropathy contributes to diabetic ED, and angiopathy (blood vessel pathology) seems to be of secondary importance. Their results also showed that poor diabetes control is associated with diabetic ED.

In the Massachusetts Male Aging Study, the age-adjusted probability of complete ED was three times greater in men being treated for diabetes than in men without diabetes (Feldman, Goldstein, et al., 1994). The prevalence of ED in men with diabetes increases with age, from 9 percent at ages 20 to 29 years to 95 percent at ages above 70 years (Vinik & Richardson, 1998), and prevalence increases with duration, poor control, and complications of diabetes (such as macrovascular and microvascular disease and neuropathies) (Fedele et al., 1998).

Schiavi, Stimmel, et al. (1993) studied forty men with diabetes but no other illnesses who were taking antidiabetic drugs, comparing them with a group of age-matched men without diabetes. The men with diabetes had more (but *not* statistically significantly more) sexuality-related difficulties, including ED on attempts at intercourse and with masturbation, sexual desire disorders, diminished frequency of intercourse, premature ejaculation, and diminished sexual satisfaction.

Although treatment with phosphodiesterase-5 (PDE-5) inhibitors (such as Viagra) is less effective in treating diabetic ED than nondiabetic ED, the convenience of a PDE-5 inhibitor that can be taken orally has popularized this treatment in a large number of men with diabetes and ED. Goldstein, Young, et al. (2003) tested vardenafil (Levitra), a newer PDE-5 inhibitor, for men with diabetes and ED. Vardenafil is more selective for PDE-5 and more biochemically potent than sildenafil-based drugs such as Viagra, both in the test tube (*in vitro*, "in glass") and in animals and humans (*in vivo,* "in life"). They found that vardenafil significantly improved erectile function in terms of successful penetration and successful intercourse, by comparison with placebo. The drug was well tolerated and had only mild adverse effects such as headache, flushing, and nasal congestion (rhinitis).

The capacity for orgasm and ejaculation usually continues even when erection is absent. In a study of 80 men with diabetes, 27 reported ED and 25 noted a decrease in sexual desire, while only 5 described any orgasmic disorder (Jensen, 1981). Of those 5 individuals, 3 had premature ejaculation and 2 had delayed ejaculation. Men with diabetes frequently experience retrograde ejaculation (Faerman, Jadzinsky & Podolsky, 1980). This is due to a dysfunction of the internal bladder sphincter resulting from diabetic neuropathy.

There are fewer studies on sexual function in women with diabetes. An early study compared 125 hospitalized women with diabetes and a control group of 100 hospitalized women without diabetes, in the age range 18 to 42 years (Kolodny, 1971). Thirty-five percent of the women with diabetes were not orgasmic, although the great majority of these women (91%) had at one time been able to experience orgasm. By contrast, only 6 percent of the nondiabetic group were anorgasmic, and these were women who had never experienced orgasm. Although anorgasmia was linked with diabetes, there was no association with age, duration of diabetes, or severity of neuropathy.

The studies described above did not specify the type of diabetes of the study participants. There are two types of diabetes mellitus: insulin-dependent (IDDM—type 1) and non-insulin-dependent (NIDDM—type 2). The following studies in women did distinguish between the two.

In 1983, Schreiner-Engel reported that the different types of diabetes may have different influences on female sexual functioning. Women with NIDDM (type 2) reported more sexual problems than women with IDDM (type 1) or control subjects. The difference could not be explained by different somatic causes, and the author suggested that women diagnosed with type 2 diabetes at an older age have increased difficulty accepting changes in various areas of their lives, including their sexual lives.

A "structured interview study" (personal interviews in which the researcher asks open-ended questions) of 42 women with IDDM, compared with both matched (nondiabetic) controls and the results from the Swedish 1986 national sex survey study, found that 26 percent of the women with IDDM had decreased sexual desire, 22 percent had decreased vaginal lubrication, and 10 percent had a decreased capacity to experience orgasm (Hulter, Berne & Lundberg, 1998). The overall level of sexual disorders in the women with IDDM was 40 percent. Among the age-matched controls without diabetes or any neurological disease, only 7 percent of the women reported any kind of sexual disorder.

Two psychophysiological studies of vaginal lubrication in women with diabetes have been published. In one, no difference was found compared with controls (Slob et al., 1990), but in the other, women with diabetes experienced significantly less physiological arousal to erotic stimuli than controls (Wincze, Albert & Bansal, 1993).

In the study of women with IDDM described above, impaired subjective vulvar sensitivity was more frequent in the women with

IDDM than in the controls (Hulter, Berne & Lundberg, 1998). The women with IDDM also showed significantly higher vibration perception thresholds (i.e., less sensitivity), measured on the hands and on the clitoris, than the controls. Other findings correlated with sexual disorders in the women with IDDM included constipation, reduced foot perspiration, and incontinence. The authors concluded that this indicates autonomic polyneuropathy (multiple disorders of the autonomic [involuntary] nervous system) as an important mechanism underlying sexual disorders in women with diabetes.

Additional possible complications of diabetes in women include vascular damage to small blood vessels as well as larger arteries, which may result in decreased blood flow to the clitoris and other erectile tissue. Clearly, more research is needed in this area.

Multiple Sclerosis

Multiple sclerosis (MS) is the most common chronic neurological disease of young adults, producing chronic and recurring neurological disability. Most researchers consider MS to be an autoimmune disease. It has a prevalence of 30 to 80 cases per 100,000 population in the northern United States and Canada, by contrast with a rate of about 5 per 100,000 in Japan, Asia, and Africa. Its slow (years-long) course involves widespread disturbances of neurological function, including motor, sensory, and gait abnormalities and impaired bladder, bowel, and sexual function (Noble, 1996).

Among men with MS, 26 to 75 percent have erectile difficulties, the prevalence depending on age, duration of MS, and severity of MS symptoms (Goldstein, Siroky, et al., 1982; Minderhoud et al., 1984). In addition, there are reports of disorders of orgasm and ejaculation, including premature ejaculation, with or without erectile problems (Schover et al., 1988).

Stenager et al. (1990) reported on the presence of sexual problems in 33 of 52 Swedish men with MS. Complaints included ED, premature ejaculation, changed sensation in the penis, reduced sexual desire, and orgasmic disorders. Only 45 percent of those who experienced changes in sexual activity or response claimed to be concerned about the change.

Another study included 381 individuals (144 men, mean age 47 years; 237 women, mean age 44 years) with MS and 291 individuals (101 men, mean age 51 years; 190 women, mean age 44 years) from the general population without MS (McCabe, 2002). Men with MS had a significantly higher frequency of premature ejaculation, failure to ejaculate, lack of sexual interest, erectile problems, difficulty masturbating, and numbness or lack of sensation than men without MS. Women with MS differed significantly from the women without MS only in their levels of masturbation and numbness or lack of sensation. A strong association was found between sexual satisfaction and relationship satisfaction for all respondents. Sexual difficulties were associated with relationship dissatisfaction only for the women with MS and not for other participants.

The findings described above are not universal, however. Barrett (1999) quotes a man writing about his difficulty in experiencing erection and orgasm: "We go to bed now, knowing that perhaps it will work but probably it will not. A feeling of alienation has entered my consciousness toward this act, which can be such an incredibly beautiful experience. I know of ways whereby my sexual feelings might be further heightened, but I rarely discuss these techniques with my wife. She always takes my failure as her failure . . . My wife's anxiety concerning my situation makes my awareness of the frustration all the more acute, and we are losing our spontaneity and exuberance."

Hulter and Lundberg (1995) reported on the changes in sexual function of women as the symptoms of MS progressed. In a group of 47 women with advanced MS, 60 percent reported decreased sexual desire, 36 percent reported decreased vaginal lubrication, and 40 percent reported diminished orgasmic experiences during the course of their disease. Sensory decrease in the genital area was experienced by 62 percent of the women, and 77 percent had weakness of their pelvic muscles. A review of other studies found that 29 to 86 percent of women with MS reported a reduced interest in sex; 43 to 62 percent, reduced sensation; 12 to 40 percent, vaginal dryness; 6 to 40 percent, dyspareunia; and 24 to 58 percent, reduced orgasmic experience (Ghezzi, 1999). Sexual problems may also occur in early and mild cases of MS (Lundberg, 1981).

Sexual disorders in MS are correlated with bladder and bowel dysfunction (Lundberg, 1981; Ghezzi, 1999). Sexual disorders are also correlated with lesions in the pons, a region of the brainstem that connects the cerebellum with the rest of the brain, as detected by MRI (magnetic resonance imaging) scans (Zivadinov et al., 2003). Brainstem lesions seen on MRI seem to be of particular importance in cases of anorgasmia (Barak et al., 1996).

In a sample of 14 women with MS, women's claims of difficult or no orgasm were significantly associated with abnormalities in or absence of cortical evoked potentials (a measure of neural response in the cerebral cortex) in response to pudendal nerve electrical stimulation (Yang et al., 2000). This response is indicative of a problem with the sensory pathway from the clitoris to the brain (Yang et al., 2000; DasGupta, Kanabar & Fowler, 2002). Lundberg (2005) recommends that women with MS compensate for this loss by applying more stimulation to the anterior vaginal wall (the area of the G spot).

In a study testing the efficacy of sildenafil for women with sexual disorders related to MS, the authors concluded that this drug is unlikely to help all female patients with neurogenic sexual disorders (DasGupta et al., 2004).

Parkinson's Disease

Parkinson's disease (PD) is one of the most common neurological diseases of advancing age. It is characterized by a pathological loss of dopamine-producing (dopaminergic) neurons in the brain. It has its greatest incidence in the fifth and sixth decades of life, but can occur at any age. Many researchers have speculated that PD is caused by a toxic or infectious environmental agent; however, no conclusive evidence has been found. The classic symptoms described by James Parkinson in 1817 remain the standard of diagnosis for this disease (Noble, 1996). The most common symptom is tremor, frequently in the hands as they lie at rest in the lap. Muscle rigidity is another major manifestation of PD. Resistance to both flexion and extension occurs evenly throughout the range of motion at both distal (e.g., wrist) and proximal (e.g., shoulder) joints. Other symptoms include delays in initiating movement (bradykinesia), difficulty reaching a target with a single continuous movement (hypokinesia), and delayed reaction times, especially for planned movements. Neuropsychological abnormalities are evident, ranging from slowed cognitive processing to frank dementia, especially in older individuals, and depression also occurs with increased frequency in people with PD (Noble, 1996).

The ability to detect fine differences in surface texture is significantly impaired in persons with PD compared with healthy volunteer subjects (Weder et al., 2000). There is a parallel between PD and the syndrome known as "sensory neglect" in laboratory rats, which is produced by lesions in the lateral hypothalamus that cut the pathways of dopamine-containing neurons. The syndrome is characterized by deficits in orienting to somatosensory stimuli (Marshall, Turner & Teitelbaum, 1971).

Bladder, bowel, and sexual disorders are prominent in people with PD compared with control subjects of the same age (Sakakibara et al., 2001). One study found that the rate of sexual disorders was not higher in people with more severe PD (Koller et al., 1990). Eighty percent of the women in the study had less frequent sexual activity than before being diagnosed with PD. The authors also found a decrease in sexual interest in 71 percent and a decrease in sexual drive in 62 percent of the women with PD. In addition, vaginal dryness was noted by 38 percent of the women, and a similar percentage reported anorgasmia. These results were also supported by the findings in a sample of younger women (Wermuth & Stenager, 1995).

Bronner et al. (2004) investigated the sexual function of 75 people with PD (32 women, 43 men) before (premorbid) and after onset of the disease. Women reported difficulties with arousal (88%), orgasm (75%), low sexual desire (57%), and sexual dissatisfaction (65%). Men reported erectile dysfunction (68%), sexual dissatisfaction (65%), premature ejaculation (41%), and difficulties experiencing orgasm (40%). For 23 percent of the men and 22 percent of the women there was a relationship between premorbid sexual disorders and cessation of sexual activity during the course of the disease.

Treatment of PD with dopaminergic drugs, which mimic or increase the effects of dopamine-producing neurons, can produce an increase in sexual function. For example, Uitti et al. (1989) described the case of a woman, 55 years of age, unmarried, and a virgin. Soon after being diagnosed with PD and beginning to take the dopaminergic drug combination levodopa/carbidopa, 500 mg/50 mg daily, "She became preoccupied with her genital region. She would frequently lie on her back and make rhythmic up and down movements similar to sexual intercourse, at times uncovering her genitalia during such episodes. After lowering the dose, her sexual preoccupation gradually declined."

In a study of 10 men with PD and erectile dysfunction who received 50 to 100 mg of sildenafil (Viagra) to use in eight sexual encounters over a two-month period, the men reported significant improvements in overall sexual satisfaction, sexual desire, ability to experience erection, ability to maintain erection, and ability to experience orgasm (Zesiewicz, Heilal & Hauser, 2001). However, this was an "open-label" study, meaning that both researchers and subjects knew which drug the subjects were taking. This differs from a "double-blind" study, in which neither the researchers nor the subjects know which drug (or placebo) any subject is taking until its identity is revealed at the end of the study; this procedure avoids bias in the evaluation of results. Zesiewicz, Heilal, and Hauser did not report whether or not a placebo was administered in another group to contrast with the experimental drug effect.

7

How Aging Affects Orgasm

If older adults are reasonably healthy and have an interesting and interested partner, they should be able to enjoy sexual relationships into very old age (Scura & Whipple, 1995). The stigmas, myths, and negative societal beliefs about older adults, such as their being incapable of enjoying sexual activity or not being interested in sex, are often accepted even by older people (Woods, 1984). They may then incorporate the judgment of society and think of themselves as "dirty old men" or "dirty old women" when they have sexual desires or relationships (Whipple & Scura, 1989). But perhaps they should more appropriately perceive themselves, and be perceived by others, as senior citizens who continue to be sexy.

A workbook by Brick and Lunquist (2003) gives excellent suggestions on sexuality education for men and women in mid- and later life. The authors present eight basic principles on sexuality and later life: "(1) Sexuality is a positive, life-affirming force; (2) Older adults deserve respect; (3) Older adults vary in their comfort with sexual language; (4) Older adults are capable of writing new sexual scripts that invigorate their sexual journey; (5) Older adults have many 'lessons' to share and learn from each other; (6) Older adults deserve accurate and explicit information and resources for additional discovery; (7) Gay, lesbian, bisexual and transgender individuals must be acknowledged; and (8) Flexible role behavior is fundamental to personal and sexual health."

Research has demonstrated that the overall best predictor of sexual activity in later life is the level of sexual activity in midlife (Knowlton, 2000). Most studies on sexuality and aging do not specifically address orgasm. Dennerstein et al. (1999) studied the factors affecting women's sexual functioning in the midlife years. Their sample included 428 Australian-born women who were 45 to 55 years of age when the study began. Annual assessments were made for six years. The authors reported that several factors significantly diminished with time (years in the study): feelings for partner, sexual responsivity, frequency of sexual activities, and libido. Other factors increased significantly over the course of the study: vaginal dryness and dyspareunia (vaginal pain during intercourse) and partner problems.

An analysis of data from the Massachusetts Male Aging Study, based on 1,085 middle-aged and older men, found that over a nine-year period, all domains of sexual function showed significant longitudinal changes, except frequency of ejaculation with masturbation, which showed no change (Araujo, Mohr & McKinlay, 2004). Decline in sexual function became more pronounced with increasing age. For example, frequency of sexual intercourse or activity decreased by less than once per month for men in their forties, by twice per month for those in their fifties, and by three times per month for those in their sixties. The number of erections per month declined by three, nine, and thirteen for men in their forties, fifties, and sixties, respectively. No mention was made in the study of orgasmic frequency, although it is likely that many men equated ejaculation with orgasm.

What Surveys Reveal

A 1999 survey by the AARP (American Association of Retired Persons) on sexual attitudes and behavior found the incidence of widowhood was 50 percent among women aged 60 to 74 years and 80 percent among older women; by contrast, only 20 percent of men aged 75 or older were widowed (Jacoby, 1999). Lack of a partner can limit sexual expression, and many of the women without partners lamented that they felt deprived of intimate kisses and hugs.

Another finding of the AARP study was that with aging, the increasing incidence of illness and progression of chronic diseases can adversely affect sexual functioning. Many pharmacological treatments for medical problems have detrimental effects on sexual function. The elderly are the largest group of consumers of prescription and nonprescription medications (Jacoby, 1999), and thus polypharmacia—the use of multiple medicines—may have a deleterious effect on sexual function in this age group.

Survey respondents also reported psychosocial issues that can affect sexual activity in later years. Changes in roles and finances after retirement, anxiety and depression related to age-associated losses and transitions, and personal, religious, and moral beliefs on sexuality in later life all affected sexual function (Jacoby, 1999).

The most recent AARP study, conducted in 2004, surveyed 1,683 adults aged 45 and older and measured attitudes and other factors affecting their sexuality and quality of life (Jacoby, 2005). One finding was that the number of men who tried treatments to enhance their potency had doubled since the earlier survey, from 10 percent in 1999 to 22 percent in 2005. The majority (69%) of these men said the treatments increased their sexual satisfaction. Women in all age groups reported that their own sexual satisfaction was enhanced by their partners' use of the drugs.

An important finding in the AARP's 2004 study was that because the potency-enhancing drugs do not affect desire, 42 percent of men who tried them stopped using them. About half of those who stopped taking the drugs said they found them to be ineffective (Jacoby, 2005). These drugs (e.g., sildenafil) depend on sexual arousal to be effective, because they act by augmenting the effects of neurotransmitters released into the penis in response to sexual arousal, thereby augmenting penile erection. In the absence of sexual arousal, there is no release of these neurotransmitters and consequently no augmentable substrate, in which case the drug has no effect on erection. These drugs have no impact on emotional problems that may be affecting sexual functioning, so underlying relationship problems can play a significant role in nonresponse to these drugs.

Nearly a third more women in the 2004 study than in the 1999 study reported masturbating. In 2004, nearly half the women in the 45 to 49 years age range reported masturbating, and 20 percent of the women aged 70 and older said they do so. "A majority of all women—even those 70-plus—told AARP that sexual self-stimulation is an important part of sexual pleasure at any age" (Jacoby, 2005).

A satisfying sexual relationship was reported by 56 percent of individuals 45 years of age and older, but it was not their highest priority. "Good spirits, good health, close ties with friends and family, financial security, spiritual well-being, and a good relationship with a partner were all rated as more important than a fulfilling sexual connection" (Jacoby, 2005). A majority (63%) of men and women with a partner described themselves as either "extremely satisfied" or "somewhat satisfied" with their sexual lives. A separate study showed that in recent years, the three main items women associated with satisfying sex were: feeling close to a partner before sex, emotional closeness after sexual activity, and feeling loved (Ellison, 2000).

As Women Age

Some studies have suggested that up to 50 percent of women notice some common sexual changes as they enter menopause naturally (Sherwin, 1993; Bachmann, 1995). Levine (1998) summarized them as follows: "(1) Slowness of vaginal lubrication; (2) Diminished volume of vaginal lubrication, occasionally to the point of painful intercourse for the woman or man; (3) Less erotic response to vulvar, clitoral, breast and nipple stimulation; (4) More difficulty focusing on the tactile sensations that previously created a state of sexual arousal; (5) Diminished drive or an increased freedom from the feeling that sex with a partner or masturbation is necessary to restore comfort; and (6) Fewer sexual fantasies and preoccupations."

Bachmann and Leiblum (2004), in a study of menopause, reported that orgasm may become shorter in duration and less intense with advancing age. Vaginal dryness, which can lead to dyspareunia, is common in older women. It can be remedied by allowing more time for the woman to become aroused and by using water-based lubricants, available over-the-counter. Topical vaginal estrogen, in cream, ring, or suppository form, may also help to alleviate dyspareunia.

The endocrine basis for menopause is not well understood. In the phase between regular menstrual cycles and menopause, ovarian production rates for individual hormones diminish to different degrees. Production of the estrogens, estradiol and estrone, decreases approximately 85 and 58 percent, respectively. Ovarian production of the androgens, androstenedione and testosterone, decreases approximately 67 and 29 percent, respectively. Progesterone production decreases dramatically—99 percent (Longcope, Jaffee & Griffing, 1981; Levine, 1998).

A major consequence of this age-related decrease in ovarian hormone production is a decrease in pelvic blood flow. As women enter their sixties, pubic hair becomes thinner and coarser, the labia majora and minora and the clitoris decrease in fullness, and fewer women exhibit color changes of the labia on sexual arousal (Masters & Johnson, 1966). The vaginal surface flattens and thins, the depth of the vagina decreases, and the walls lose their elasticity. The biochemical environment becomes less acidic, which creates a shift in the vaginal flora, and the uterus returns to its prepubertal size. Orgasmic contractions of the vagina still occur at age 60, but contractions of the rectum do not seem to occur as in premenopausal women. Common symptoms that are related to these changes are vulvodynia (chronic discomfort of the vulva perceived as burning, stinging, or irritation), dyspareunia, post-intercourse spotting of blood, post-intercourse urinary tract infections, and vaginitis (Levine, 1998).

As Men Age

Men between the ages of 40 and 50 years generally experience a decline in sexual drive and loss of erectile intensity. The most common age at which men seek help for ED is in their late fifties (Levine, 1998). Schiavi, Schreiner-Engel, et al. (1990) noted a significant decline in the frequency and duration of nocturnal erections in men between 45 to 54 and 55 to 65 years of age. It seems, based on their study, that the male sexual system declines during those years. In 60-year-olds, the degree of erectile unreliability is 30 percent (Feldman, Goldstein, et al., 1994). A study of ED among 1,290 men between the ages of 40 and 70 years found that 10 percent were completely, 25 percent moderately, and 17 percent minimally impaired (Mulhall & Goldstein, 1996).

One of the benefits reported for older men is that some are able to attain good ejaculatory control for the first time in their lives. Another benefit is that even though erections may not be as reliably firm, they are adequate for intercourse. Perhaps the diminished turgidity decreases vaginal friction in heterosexual couples and requires a lesser degree of vaginal lubrication (Levine, 1998).

Whipple (2005), in summarizing the literature, described several changes that may occur in men as they age: (1) it takes a longer time and more direct stimulation for the penis to become erect; (2) the erection becomes less firm; (3) the amount of semen is reduced and the force of ejaculation diminishes; (4) there is less urge to ejaculate; and (5) the refractory period lengthens.

Testosterone levels are highest in the morning. Some people find that scheduling sexual activity in the morning is preferable to other times of day (Levy, 2002). It is not known whether testosterone plays a role in this preference or is just one among many physiological factors that fluctuate in a daily rhythm and influence sexual vigor (e.g., adrenal corticoid hormone levels are also highest in the morning).

Several pharmaceutical agents can be helpful for older men. Although Viagra, Cialis, and Levitra are frequently prescribed, they are not a panacea. Approximately half the men who try one of these medications discontinue its use by the end of one year. Difficulties include inadequate education on what to expect from the drugs and inadequate communication with partners (Goldstein, 2002). Whipple (2000, 2002a) reported that if the partner participates during the man's medical assessment and helps to decide on treatment options, there is better compliance with the treatment plan.

Safer Sex for Seniors

In many societies, older adults are often considered to be asexual and not at risk for HIV/AIDS. They are usually excluded from donating blood, and thus the routine HIV screening this provides, due to age limits imposed by blood banks (Scura & Whipple, 1995). Older adults also tend to believe their behavior does not place them at risk for HIV infection. Few older adults consider using condoms, as they are not concerned with birth control. However, in the United States, AIDS is diagnosed in more people 50 years of age and older than in those 24 and younger. In 1991, more than 10 percent of U.S. AIDS cases involved people 50 years of age and older (Centers for Disease Contol, 1991). By 2003, 23 percent of the diagnoses of HIV were in people over 45 years of age, and 30 percent of people in this age group who were diagnosed with AIDS had been infected through heterosexual intercourse (Gottesman, 2005).

Dealing with the Aging Process

So what can older persons do to maintain their sexual health? Sex, like walking, does not require the stamina of a marathon runner, but it does require reasonably good health. A basic perspective is "use it or lose it." Although the physiological mechanisms are unclear, prolonged abstinence from sexual activity can apparently promote ED, and women who are sexually active (with partner or self) after menopause have better vaginal lubrication and elasticity of vaginal tissues than women who are sexually inactive. It is also beneficial to follow a balanced, low-fat diet and to exercise regularly. Fitness enhances self-image and improves vigor. Men who smoke heavily are more likely to have ED than nonsmokers, and people who smoke are at an increased risk of arteriosclerosis (hardening of the arteries), which can cause reduced blood flow and thus ED or reduced vaginal lubrication. Chronic alcohol and drug abuse creates psychological and neurological problems related to ED, especially if the liver is damaged (Whipple, 2005).

Based on interviews with older couples about sensuality and sexuality, Whipple (2005) observed that both men and women need more time to become sexually aroused than when they were younger. Older couples may have to increase sex play and may discover that they can satisfy their sensual and sexual desire with fondling, caressing, and kissing ("outercourse"). They should be aware that illness or fatigue may impair sexual response of one or both partners and that alcohol, tranquilizers, and many other medications can reduce sexual drive and alter the ability to respond sexually. As many options for sensual and sexual pleasure are available to older adults as to younger adults. Masturbation, fantasy, and "outercourse" are always options.

8

Pleasure and Satisfaction with and without Orgasm

Is orgasm the ultimate goal of sexual interactions? Not everyone thinks so. We view sexuality and sexual expression holistically, an approach that does not focus on orgasm as a goal. Cultural views of sexuality have evolved from an earlier, predominantly procreational role for sexual activity to the current view in many societies that sexuality entails not only procreation but also pleasure and physical and mental health and well-being. Sexuality and sensuality can enhance the quality of life, foster personal growth, and contribute to human fulfillment (Whipple & Gick, 1980). When *sexuality* is viewed holistically, it refers to the totality of a being, not just to the genitals and their functions. Sexuality includes all the qualities that comprise a person—biological, psychological, emotional, social, cultural, and spiritual. And a person has the capacity to express his or her sexuality in any or all of these areas, without necessarily involving the genitals (Whipple, 1987).

There are two commonly held views about sexual expression (Timmers, Sinclair & James, 1976). The most prevalent sees sexuality as linear and goal-oriented, much like climbing a flight of stairs. The first step is touching, the next is kissing, followed by caressing, then, for heterosexual couples, vagina-penis contact followed by intercourse, and then the top step—orgasm. If the sexual experience does not lead to orgasm, one or both partners feel unsatisfied (Whipple, 1987). An alternative view of sexual expression is pleasure-oriented, conceptualized as a circle, with each form of expression on the perimeter of the circle considered an end in itself. Whether the experience involves kissing, holding hands, cuddling, oral sex, or other expressions, each is an end in itself and each is satisfying to the person or the couple. Stereotypically (and acknowledging the limitations of the stereotypes), women tend to be more pleasure-oriented and men tend to be more goal-oriented (Whipple, 1987).

It is important to realize that historically, satisfaction has not been considered as an outcome criterion for sexual function and sexual disorders. In 1998, an international multidisciplinary group of experts on female sexuality assembled and drafted a classification system that built on and moved beyond the frameworks of the Diagnostic and Statistical Manual of Mental Disorders (DSM-IV; American Psychiatric Association, 1994) and the International Statistical Classification of Diseases and Related Health Problems (ICD-10; World Health Organization, 1992). The group developed the Consensus-Based Classification of Female Sexual Dysfunction (CCFSD; Basson, Berman, et al., 2000). In spite of its notable advances over the existing classification systems (by including both organic and psychogenic dysfunction and personal distress, and moving away from the heterosexual bias found in other classification systems), the CCFSD still has significant flaws. One of the major flaws is that "satisfaction" is not included as an outcome criterion, despite its inclusion being recommended by a majority of the members of the 1998 multidisciplinary group.

As many of the experts pointed out, in thirty-seven published commentaries in a special issue of the *Journal of Sex and Marital Therapy* in 2001, there are problems with the CCFSD in that it is based on a triphasic pattern of sexual function. Although in widespread use, this model is based on the male, linear model of desire, arousal, and orgasm (in that order); this may not describe the sexual experience of women. Women can experience sexual arousal, orgasm, and satisfaction without desire, and they can experience desire, arousal, and satisfaction without orgasm. A woman who has sexual satisfaction but does not go through all the linear phases of the sexual response model should not be construed as having a sexual disorder (Sugrue & Whipple, 2001).

Most health care providers who work with women and their sexual concerns report that women's sexual experience is more complex than having or not having an orgasm or the presence or absence of vaginal lubrication. Women's sexual experiences encompass self-esteem, body image, relationship factors, pleasure, satisfaction, and many other variables. Health care providers and consumers would be well advised to recognize and acknowledge the variety of ways in which women experience sexual and sensual pleasure (Whipple, 2002b).

Sugrue and Whipple (2001) developed a classification system that is based on a biopsychosocial understanding of women's sexual experience, does not "medicalize" a woman's sexuality, and reflects a woman's—not a man's—experience of sexuality. This classification system includes:

—Capacity to experience pleasure and satisfaction independent of the occurrence of orgasm
—Desire for, or receptivity to, the experience of sexual pleasure and satisfaction

—Physical capability of responding to stimulation (vasocongestion) without pain or discomfort
—Capability of experiencing orgasm under suitable circumstances

If these or similar descriptors were viewed as characteristic of normative sexual function, then the persistent absence or modification of any of them would constitute a sexual dissatisfaction (Sugrue & Whipple, 2001; Whipple, 2002b).

Some men have also reported that pleasure and satisfaction are important to them and that they do not necessarily need to experience an orgasm for pleasure and satisfaction (Whipple, 2002b).

One of the reports based on the Second International Consultation on Erectile and Sexual Dysfunctions (later known as the International Consultation on Sexual Medicine), held in Paris in 2003, contains the statement that "sexual activity in women involves interest and motivation, ability to become aroused and experience orgasm, the pleasure of the experience and subsequent personal satisfaction" (S. R. Davis, Guay, et al., 2004). Thus, the importance of pleasure and satisfaction in sexual relations seems to be gaining increasing recognition and appreciation.

9

The Nervous System Connection

In the course of evolution, two integrative systems developed in vertebrates: the nervous system for rapid communication and the endocrine system for messages developing more slowly but lasting longer (in some cases for life). Both systems use chemical signals: the nervous system uses neurotransmitters and neuromodulators; the endocrine system uses hormones. In this chapter our focus is on the nervous system.

Our knowledge of the role of neurotransmitters in the expression of orgasm is mainly derived from the use of psychotropic drugs, which affect the nervous system. Understanding how the nervous system operates is key to understanding how orgasm and pharmaceutical or other drugs interact.

Neurons and Neurotransmitters

Neurotransmitters are synthesized by neurons (nerve cells) in the central nervous system (CNS; the brain and spinal cord) and in ganglia (groups of neuron cell bodies) located outside the CNS. (A neuron consists of a cell body and a long extension called an axon.) Examples of ganglia are those of the autonomic nervous system, also known as the "involuntary" nervous system, that innervate (i.e., send axons to) the viscera and control heart rate, blood pressure, gastric secretion, sweating, epinephrine (adrenaline) secretion, and so forth.

The sites of action of neurotransmitters are highly specific and localized. Thus, a ganglionic neuron that sends its axon to the heart delivers its chemical message to cardiac muscle cells but not to cells of the liver. Similarly, within the brain, a neuron in the thalamus that sends its axon to neurons in the cerebral cortex modulates the activity of cortical neurons but not neurons in other parts of the brain. This restriction of the message is due not only to the established "wiring" of the neural circuits but also to the limited diffusion of the chemical signals: the neurotransmitters. Neurotransmitters are the chemical messengers between neurons. They are released from the axon terminals of one neuron into a specialized structure called the "synapse," the microscopic space between communicating neurons. On the other, receiving, side of the synapse is the next neuron in the neural pathway, which increases or decreases its activity in response to the neurotransmitter released into the synapse (Bloom, 2001).

Neurons can both send and receive messages through the synapses. When stimulated or spontaneously activated, neurons send electrochemical impulses (known as "action potentials") from one part of the neuron along the axon to the axon terminals. Usually, these impulses do not jump to the neighboring cell, but instead release chemicals—the neurotransmitters—from the axon terminals into the synapse. The neurotransmitter molecules released by this "presynaptic" neuron may stimulate or inhibit the receiving neuron(s) on the other side of the synapse, the "postsynaptic" neuron(s). This occurs mainly through the neurotransmitter's effect on the receptive antennae-like filaments (dendrites) on the postsynaptic neuron or directly on the cell body of the postsynaptic neuron. Some axon terminals release their neurotransmitters onto nonneuronal cells via specialized nonsynaptic contact zones, producing such effects as muscle cell contraction or salivary gland secretion (Davenport, 1991).

Action potentials are converted to chemical signals by a process termed "excitation-secretion coupling." Neurotransmitters are typically small molecules such as amines (epinephrine, serotonin) or amino acids (glycine, GABA). They are stored inside the axon terminals in microscopic synaptic vesicles. When the axon is activated (i.e., carries an electrochemical impulse), the vesicles merge with the cell membrane of the axon terminal, open up to the outside of the neuron, and release their contents—hundreds of neurotransmitter molecules—into the synapse. The released neurotransmitter molecules find and bind to specialized receptors, complex protein structures situated at specific sites on the postsynaptic neurons (Kenakin, Bond & Bonner, 1992). The binding of a neurotransmitter to a specialized receptor embedded in the membrane of the postsynaptic neuron triggers a series of rapid chemical reactions in that neuron. This results in either the generation of a new nerve impulse in the postsynaptic neuron or a cascade of biochemical events that change the excitability and capacity of the postsynaptic neuron to respond to other incoming stimuli for variable periods of time (Girault & Greengard, 1999). (Note that neurotransmitters stimulate or inhibit the action potential–producing *activity* of neurons, and neuromodulators change the *responsiveness* of neurons to the neurotransmitters.)

The neurotransmitter is also termed a "first messenger." As a consequence of the first messenger binding to the receptor in the membrane of the postsynaptic neuron, a chemical linkage activates another chemical, the "second messenger." In all, this activation requires the sequential involvement of four molecules: (1) the neurotransmitter, or first messenger; (2) the specific neurotransmitter receptor; (3) a "transmembrane-connecting protein," termed the G protein; and (4) an enzyme inside the postsynaptic neuron that synthesizes the second messenger. Well-characterized second messengers are the cyclic nucleotides cyclic AMP (cAMP) and cyclic GMP (cGMP).

The second messenger triggers a cascade of cellular events that involves enzymes called "kinases," which phosphorylate (add a phosphate group to) other proteins. Phosphorylation can alter various processes in the neuron, such as the movement of ions ("ion flux") in the membrane or the rate of synthesis of certain proteins (enzymes, receptors) by the cell's genetic machinery (the genome) (Girault & Greengard, 1999). "Excitatory" neurotransmitters (e.g., glutamate) stimulate the postsynaptic neuron to increase its rate of generation of action potentials, whereas "inhibitory" neurotransmitters (e.g., glycine and GABA) decrease the rate of generation of action potentials. Other neurotransmitters (e.g., epinephrine, serotonin, dopamine) may induce either excitation or inhibition of the postsynaptic neuron, depending on where in the nervous system they are released. This apparently paradoxical effect—the same chemical producing opposite effects—is possible because neurotransmitters act on a variety of receptors. Each type of receptor interprets the message in a different way, depending on its relationship with that cell's response mechanisms (or "effector" mechanisms), such as the specific second messengers it produces.

Neurons do not communicate only with other, postsynaptic neurons. They also modulate their own activity as their own neurotransmitters act on "autoreceptors" located in their own membrane. Autoreceptors, when activated by the cell's own neurotransmitter, trigger different responses depending on where on the presynaptic membrane they are located. If located in the membrane of the cell body or its dendrites (somatodendritic autoreceptors), the response may be a decrease in the neuron's firing rate. If located in the membrane of the axon terminals (axonic autoreceptors), the release of neurotransmitters from the terminals is inhibited.

Receptors, Neurotransmitters, and Drugs

The number of different neurotransmitters and neuromodulators identified in the nervous system has increased dramatically in the past thirty years; current estimates suggest there are at least fifty types. Estimates of the number of receptor types on which neurotransmitters act have increased even more (J. R. Cooper, Bloom & Roth, 2003). Neurotransmitters do not act on only a single type of receptor; all the neurotransmitters have the capacity to interact with several distinct subtypes of receptors. Some neurotransmitters interact with only two receptor subtypes, as in the case of the inhibitory neurotransmitter glycine, while others, such as serotonin and norepinephrine, can act on more than ten receptor subtypes. Receptors are also the sites of psychotropic drug action.

Despite the capacity of neurotransmitters to activate several different receptor subtypes, the interaction between each neurotransmitter and receptor is highly specific. Paul Ehrlich, one of the founders of pharmacology, compared the interaction of a drug with its receptor to that of a key and a lock, a metaphor that reflects the high selectivity of receptors for the ligands that bind to them. (*Ligand* is the general term for a substance—a neurotransmitter or drug—that binds to a receptor.) Just as slight changes in the shape of a key prevent it from opening the lock, slight changes in the chemical structure of a ligand can prevent it from exerting its action on a receptor.

In some cases, the structural change in the ligand does not prevent it from *binding* with the receptor, but it blocks the subsequent cascade of cellular events that would otherwise be triggered by the binding. Indeed, in such cases, the ligand acts as a receptor blocker (or receptor antagonist), in that it does not exert the typical neurotransmitter effect but blocks the ability of other, active neurotransmitter molecules to bind to the receptor. To extend the metaphor, it is as if a key breaks off and gets stuck in the lock, preventing an intact key from entering and opening the lock. The broken-off key is the receptor blocker.

Neurotransmitters are like master keys capable of unlocking each of the various receptor subtype locks. Drugs can mimic the action of the neurotransmitter on some receptor subtypes but not others. Particularly useful for pharmacological studies are drugs that *duplicate* the effect of a neurotransmitter on only a single receptor subtype (a drug acting as a "specific agonist") or *selectively block* the effect of the neurotransmitter on a specific receptor subtype (a "specific antagonist"). Such drugs are like a submaster key, having an effect at only one of the receptor subtypes (Neubig & Thomsen, 1989).

Chemical transmission in the nervous system, then, uses multiple neurotransmitters, each one acting through multiple receptor subtypes. According to Stahl (1999), this provides both selectivity and amplification. Thus, for example, the serotonin receptor family ("family" meaning the set of receptor subtypes that bind a particular neurotransmitter) responds selectively to a single neurotransmitter, serotonin, but receptor communication is amplified due to the presence of the great variety of serotonin receptor subtypes. Since neurotransmitters are released in a well-circumscribed anatomical space in close proximity to postsynaptic neurons that typically have only one or two specific receptor subtypes, the simultaneous activation of all receptors for a particular transmitter rarely occurs under physiological conditions.

This is in contrast to the situation when a drug is administered into the bloodstream (i.e., administered systemically), in which case all or most subtypes of a neurotransmitter receptor family are simultaneously and indiscriminately exposed to the drug. For example, some drugs that are effective in treating depression increase the concentration of serotonin in all synapses by blocking its reuptake from the synapse. (In the normal course of events, some neurotransmitter molecules are removed from the synapse via reuptake by the neuron that released them; others are destroyed by enzymes.) Under these conditions, many or all serotonin receptor subtypes may be activated and participate in the biological responses (we discuss these reuptake inhibitors again later in the chapter). Differentiation and identification of the specific receptor subtypes that participate in the production of the desired effect—such as an antidepressant effect—requires research, often using specific agonists or antagonists for ascertaining the receptor subtype responsible for the response.

There is an added complication in the analysis of drug action. As noted earlier, a neurotransmitter can exert opposite effects by acting on different receptor subtypes. The type of response obtained, whether excitation or inhibition, depends on the chemical pathway within the cell. Thus, besides classifying receptors on the basis of the neurotransmitter to which they respond, receptors are also assigned to one of two different "superfamilies," either ligand-gated ion channel receptors or G protein–linked receptors. Both types are embedded in the neuronal cell membrane, but they differ in their structure and mechanism of action.

Ligand-gated ion channel receptors are proteins that surround openings (ion channels) in the cell membrane through which ions—electrically charged molecules—can pass. The long protein chain of these receptors looks as if it is threaded, or looped, in and out through the cell membrane, like a purse string, surrounding an ion channel, hence it is termed a "transmembrane" protein. This transmembrane protein chain is the receptor. When a ligand binds to this receptor, it changes the shape of the protein, thereby dilating or constricting the size of the ion channel and permitting or blocking passage of specific ions (K^+, Na^+, Ca^{2+}, or Cl^-) into or out of the cell. Thus, the job of these receptors is to modulate (i.e., to "gate") the transport of ions through the channel. The relative concentrations of these different ions inside and outside the neuron at any moment determine the readiness of the neuron to respond to an arriving signal (synaptic input).

Receptors in this superfamily contain five different specific protein regions surrounding the ion channel. The channels open and close very rapidly (on the order of a few thousandths of a second), the ions flow in and out through the channels in the cell membrane, and the neuronal excitability changes almost immediately.

G protein–linked receptors contain seven transmembrane regions arranged in a circle to form a central core where the neurotransmitter binds. All receptors in this family are linked to a G protein, which associates with enzymes to produce second messengers (e.g., with the enzyme adenylate cyclase to form cAMP). The effects mediated by these receptors, also called "metabotropic" receptors, are much slower (seconds or even minutes) than those of the ligand-gated ion channel receptors. Activation of G protein–linked receptors generates a cascade of cellular events involving and affecting a variety of processes including both genomic and nongenomic effects, all of which have different time courses (Nestler & Duman, 1998).

Several different neurotransmitters act on both types of receptors. Thus, acetylcholine acts on a type of ligand-gated ion channel receptor known as a "nicotinic" receptor and on several subtypes of G protein–linked receptors termed "muscarinic" receptors. Serotonin acts on fourteen subtypes of receptors, thirteen associated with various types of G proteins and one, the 5-HT3 receptor, linked to a channel involved in K^+ transport across the membrane (the chemical name for serotonin is 5-hydroxytryptamine, hence its abbreviation as 5-HT).

Drugs rarely affect a single receptor subtype, or even a single family of receptors, but usually interact with several types of receptors. For example, clozapine, an "atypical" antipsychotic, interacts with at least nine types of receptors, thus making it difficult to ascribe its biological effect to a specific receptor interaction. Researchers must compare results obtained with many different drugs to attribute to a specific receptor a role in a given behavior.

The effectiveness of psychoactive drugs in treating mental diseases is largely based on their capacity to interact with receptors in the brain that are activated by neurotransmitters. Drugs (exogenous ligands) that mimic the effect of a neurotransmitter (an endogenous ligand) on a receptor are termed "agonists." Thus, morphine is an opiate agonist that produces its powerful analgesic effect through its binding to an opiate receptor. The brain makes its own opiate-like chemicals, the endorphins (a term derived from *endo*genous mo*rphin*e), which were originally defined as any *endogenous* substance that has the pharmacological properties of morphine.

Other agents also bind to the opiate receptor but fail to trigger the chain of events that would ultimately generate a behavioral response. Consequently, these agents are termed opiate *antagonists,* which shield the receptors from the opiate *agonists.* Antagonists are widely used in medical practice—for example, drugs that block the response to histamine (histamine antagonists, or antihistamines), which provide relief from the troublesome symptoms caused by allergens such as pollen or dust.

How Modulation of Neuronal Activity Affects Orgasm

Orgasm is affected by particular agonists and antagonists. For example, the drug buspirone, a specific agonist of serotonin type 1A receptors, facilitates some aspects of human sexual response, including orgasm (Norden, 1994). Yohimbine, an antagonist of the adrenergic alpha-2 receptor subtype located on the presynaptic terminals of adrenergic neurons (i.e., neurons that synthesize and use norepinephrine, which is also called noradrenaline, as their neurotransmitter), blocks the inhibitory action of norepinephrine on these autoreceptors. By blocking the inhibitory action, yohimbine increases the release of norepinephrine from the neurons. This ultimately results in an "aphrodisiac" effect (E. Hollander & McCarley, 1992).

In many cases, drugs affect sexual response via mechanisms other than a direct action on receptors. For example, amphetamine (which has been reported to facilitate the expression of orgasm) acts by releasing dopamine from neuron terminals (Kall, 1992). It is understandable, therefore, that the dopamine receptor antagonists (the "neuroleptics") such as chlorpromazine or haloperidol have the side effects of impairing various aspects of sexual response, including orgasm (Shen & Sata, 1990).

Most of the drugs that affect orgasm in men and women influence the number of neurotransmitter molecules present in the synapse or prolong the amount of time the neurotransmitter remains in the synapse. These drugs interfere with the normal biochemical processes that terminate the action of neurotransmitters, either through enzymatic breakdown or by reuptake. The enzymes that break down neurotransmitters usually have a highly specific target—that is, they degrade only certain molecules and not others. The enzyme monoamine oxidase, for example, acts only on monoamines such as norepinephrine, serotonin, and dopamine. Drugs that inhibit these degradative enzymes elicit changes in the postsynaptic neuron by prolonging the action of the neurotransmitter in the synapse, thereby exerting an indirect effect on the postsynaptic neuron.

Another family of drugs that affect neurotransmitter action are the specific reuptake inhibitors. Some neurotransmitters, including the monoamines, return to their neurons of origin immediately after being released into the synapse and are eventually reused. It is this reuptake process, clearing them from the synapse, that terminates their action. Reuptake involves a series of events, including association of the neurotransmitter with a carrier protein in the neuronal membrane that, with the help of an energy-providing enzyme system, transports the neurotransmitter back into the presynaptic neuron. Drugs that obstruct the attachment of a neurotransmitter to its carrier protein amplify the effect of the neurotransmitter by prolonging its presence in the synapse. This is the mechanism by which the selective serotonin reuptake inhibitors (SSRIs) exert their antidepressant action and contribute to anorgasmia (Mitchell & Popkin, 1983).

Some recreational drugs that enhance sexual response are reuptake inhibitors of neurotransmitters other than serotonin. For example, cocaine inhibits the reuptake of dopamine, thus prolonging its action on postsynaptic dopamine receptors (N. S. Miller & Gold, 1988). Activation of some of these dopamine receptors, specifically those classified as D2 receptors, produces mental states that some individuals perceive as improving their social interactions or enhancing their sexual behavior or their affective response to sexual stimulation.

Other drugs that influence sexual response act by prolonging the action of the second messenger, rather than that of the neurotransmitter itself. Thus, sildenafil (Viagra) facilitates penile erection by prolonging the effect of the second messenger cGMP, which is generated by the action of the neurotransmitter nitric oxide. Sildenafil acts by preventing the action of the enzyme, phosphodiesterase-5, that breaks down cGMP (Boolell et al., 1996).

10

The Neurochemistry of Orgasm

Some of the neurotransmitters involved in ejaculation are chemically identical in many different species of vertebrates, indicating that they are highly conserved from an evolutionary point of view. Across species, two neurotransmitters seem to be essential to ejaculation: dopamine and serotonin. These two neurotransmitters also seem to mediate orgasm in men and women, and sexual behavior in male and female animals.

Dopamine as a Trigger for Sexual Behavior

Dopamine plays a facilitatory role in male sexual activity, not only in humans but in other mammals and in reptiles and birds. In rodents there is extensive evidence that an increase in dopaminergic tone facilitates sexual arousal and ejaculation (Melis & Argiolas, 1995). Indeed, such effects on sexual behavior produced by drugs that act through dopamine were shown even before dopamine was established as a neurotransmitter, at a time when it was considered only as a precursor of norepinephrine. Thus, in 1957 Soulairac and Soulairac reported that amphetamine, a drug later shown to release dopamine, facilitated ejaculation in male rats, reducing by 33 percent the number of intromissions required to ejaculate and increasing by 30 percent the number of ejaculations. This finding that an increase in dopaminergic tone stimulates ejaculation has been confirmed repeatedly. Based on the number of intromissions (penile insertions into the vagina) preceding ejaculation as an index of ejaculatory "threshold," a large number of dopamine precursors, agonists, dopamine releasers, and dopamine reuptake blockers have been found to decrease this threshold. (These drugs include D-amphetamine, apomorphine, pergolide, Nn-propyl norapomorphine, lisuride, and LY 163 502.)

The impetus to study the role of dopamine in male sexual behavior in rats originated in reports that for patients with Parkinson's disease, administration of the dopamine precursor L-dopa stimulated penile erection and sexual activity (Bowers, Van Woert & Davis, 1971). A few years later, this apparent "aphrodisiac" effect was corroborated in rats (Tagliamonte, Fratta & Gessa, 1974).

For females, however, the findings on the participation of dopamine in sexual behavior are contradictory. Pharmacological observations indicate a facilitatory role of dopamine in sexual arousal and orgasm in women (Uitti et al., 1989; Shen & Sata, 1990). In female rats, dopaminergic drugs have more variable effects on estrous (mating) behavior. Reports on agents that deplete dopamine or that antagonize the effects of dopamine suggest that dopamine has mainly an inhibitory effect on lordosis (elevation of the rump and vaginal opening to facilitate penile insertion) in the rat. However, some researchers have reported facilitatory effects on lordotic behavior by the administration of low doses of dopamine agonists in female rats that are relatively unreceptive sexually (for a review, see Melis & Argiolas, 1995).

We can resolve this apparent discrepancy by recognizing that dopaminergic stimulation has two different effects in rats: stimulating locomotion and increasing sensitivity to sensory stimulation. During mating, the female rat stands rigidly immobile in the lordotic posture as the male mounts her. After intromission the female suddenly darts away from the male (Komisaruk & Diakow, 1973). Since the mating period is characterized by intense running activity in the female rat, this creates a dynamic tension between the tendency to run and the tendency to stand immobile. Perhaps it is this dynamic tension that generates the abrupt darting and stopping that characterize the female rat's mating behavior (Komisaruk, 1971). Consistent with the obligatory immobilization that allows mating, drugs that block the locomotion-stimulating action of dopamine augment the female rat's mating response (Everitt, Fuxe & Hokfelt, 1974; Herndon et al., 1978). Conversely, dopaminergic drugs stimulate locomotion, so the female does not stand still for the male's mounts and copulation is thwarted (Caggiula et al., 1979).

By contrast, dopaminergic stimulation also produces a marked increase in responsiveness to sensory stimulation, consistent with findings that blocking the dopamine system produces a syndrome of "sensory neglect" (Marshall, Turner & Teitelbaum, 1971). Thus, it is likely that response to dopaminergic stimulation depends on individual differences in the balance struck between these two effects—locomotion and sensitization. The sensitization effect may predominate over the locomotory effect and augment the female rat's mating response to the stimuli provided by the mounting male. Or the locomotion-stimulating effect may predominate and the female will not stand still to allow the male to mount. These opposing effects could account for the unpredictability of dopamine manipulation in female rats.

Given the close agreement between the results obtained in humans and rats on the action of dopaminergic drugs on male sexual behavior, we can infer that similar neural structures and circuitry are involved in both species. This is consistent with observations that the anatomical organization of the dopaminergic system, in terms of nuclei and brain connections, is very similar across mammalian species (Bjorklund & Lindvall, 1984).

Various methods have been used to study the neurotransmitter basis of sexual behavior in male rodents, such as destroying (lesioning) different components of the underlying neural system, infusing agonists or antagonists into specific brain regions, and using modern methodology such as pulse voltammetry and microdialysis to measure rapidly changing, very low levels of neurotransmitters released locally in the brain in awake animals (for reviews, see Mas et al., 1990; Pleim et al., 1990; Melis & Argiolas, 1995).

Studies performed in the rat have revealed some of the brain structures that mediate dopaminergic effects on sexual behavior. The mixed D1-D2 agonist apomorphine, when infused into the medial preoptic area of the forebrain, increased the number of ejaculations and the proportion of rats that copulated by 40 percent. Apomorphine was ineffective when infused into other forebrain regions—either the nucleus accumbens or other regions of the striatum. Dopamine receptor antagonists such as haloperidol produced the opposite effect when infused into the medial preoptic area, decreasing the number of intromissions and ejaculations by 50 percent (Melis & Argiolas, 1995). An increase in dopaminergic activity in the medial preoptic area during copulation in the rat has been detected by both microdialysis and chronoamperometry, another highly sensitive method that measures the levels and time course of specific neurotransmitters as they are released into local brain regions during behavior (A. G. Phillips, Pfaus & Blaha, 1991). Apomorphine infusions into these areas in rats elicit penile erection (Melis & Argiolas, 1993), but not when directed to other forebrain regions—the striatum or nucleus accumbens. However, a significant increase in endogenous dopaminergic activity was observed in the nucleus accumbens in copulating male rats. Researchers have interpreted this finding as being related to the rewarding effect of copulation and ejaculation that could be mediated by dopamine release from the mesolimbic pathway into the nucleus accumbens (Pleim et al., 1990; A. G. Phillips, Pfaus & Blaha, 1991).

Besides participating in neural processes that facilitate ejaculation and sexual gratification, dopamine also facilitates and even elicits penile erection in both rats and humans. This action is mediated through D2 receptors. Thus, mixed (D1-D2) or D2 agonists (e.g., apomorphine, bromocryptine, lisuride), but not D1 agonists, induce penile erections when administered systemically. Facilitation of penile erection by dopamine agonists such as apomorphine requires androgen priming, since it does not occur in castrated males. However, this response can be restored by testosterone or by the combined administration of two other steroid hormones: estradiol and dihydrotestosterone (Melis, Mauri & Argiolas, 1994).

The D2 postsynaptic receptors responsive to dopamine agonists in eliciting penile erection are located in the medial preoptic area and the paraventricular nucleus of the hypothalamus (Buijs et al., 1984; Lindvall, Bjorklund & Skagerberg, 1984).

Dopaminergic axons that project to the paraventricular nuclei originate from a small group of neurons, termed the A-14 dopaminergic group, which constitute the incertohypothalamic pathway. In rats, the axons originate in a forebrain area, the subthalamus, and project to the hypothalamus. In the paraventricular nucleus, D2 receptors are located on oxytocin-synthesizing neurons (Buijs et al., 1984; Lindvall, Bjorklund & Skagerberg, 1984).

These data show that in the male rat, dopamine plays three important roles in sexual behavior, most likely acting through anatomically distinct regions of the dopaminergic system. First, penile erection depends on an incertohypothalamic dopaminergic component (originating in the subthalamus of the forebrain) linked to the oxytocinergic system. Second, several components of copulatory behavior, such as ejaculation threshold, latency to ejaculation, and postejaculatory interval, are modulated by the medial preoptic area. Third, projections of the midbrain ventral tegmental area (area A10) to the nucleus accumbens and to the limbic cortex may be related to the rewarding aspects of ejaculation.

Although the participation of dopamine in the production of orgasm and ejaculation in men seems well supported, unresolved questions remain. For example, there is good evidence from several subprimate studies that the medial preoptic area is essential to male sexual behavior. Destruction of the medial preoptic region prevents the expression of male copulatory behavior in rats (Larsson & Ahlenius, 1999), androgen implants in this area restore behavior in castrated rats (Moralí, Larsson & Beyer, 1977; E. R. Smith, Damassa & Davidson, 1977), and in gerbils, preoptic neurons express the Fos protein (a marker of neural activation) following ejaculation (Heeb & Yahr, 2001; Simmons & Yahr, 2002). However, Holstege et al. (2003), using PET (positron emission tomography) in men, found that the medial preoptic area did not show an increase in activity during ejaculation. This apparent discrepancy between animal and human data may be due to a technicality such as limited resolving power or sensitivity of the PET methodology, but it may also be due to the process of encephalization (increasing brain complexity) by which functions integrated in subcortical structures (e.g., the medial preoptic area) in subprimate mammals are transferred to the cerebral cortex in humans.

Serotonin as a "Brake" on Sexual Behavior

The other major modulator of sexual behavior, besides dopamine, is serotonin. Serotonergic neurons are found in nine cell groups, classified as B1 to B9, located mainly in the pons and the mesencephalon of the brainstem (Frazer & Hensler, 1999). As in the case of the localization of dopaminergic neurons, the neuroanatomy of the serotonergic system in humans is comparable to that in many other mammalian species.

There is abundant pharmacological evidence indicating that the serotonergic system exerts a tonic *inhibitory* effect on male sexual behavior. Thus, drugs such as para-chlorophenylalanine (pCPA) that decrease brain serotonin levels (by interfering with serotonin synthesis) facilitate sexual behavior, particularly in noncopulators, or castrated rats treated with low doses of testosterone (for a review, see Hull, Muschamp & Sato, 2004). By contrast, raising serotonin levels in rats by administering 5-hydroxytryptophan (5-HTP), the precursor of serotonin, or injecting serotonin directly into the brain (Hillegaart, Ahlenius & Larsson, 1991) increased the number of intromissions required to achieve ejaculation and prolonged the latency to ejaculation (Ahlenius & Larsson, 1991). Consistent with these findings, in men, the use of some selective serotonin reuptake inhibitors (SSRIs), which elevate serotonin levels at the synapse, also inhibits ejaculation and probably also sexual motivation. In rats also, SSRIs such as fluoxetine were effective in inhibiting sexual behavior when administered chronically, but not acutely (Cantor, Binik & Pfaus, 1999; Mos et al., 1999).

Based on studies that used a variety of serotonin agonists, it is evidently the postsynaptic serotonin 2 receptors that mediate the inhibitory effects of this neurotransmitter on male sexual behavior (Haensel, Rowland & Slob, 1995). Consistent with these findings, stimulation of (presynaptic, inhibitory) serotonin 1A receptors in testosterone-treated male rats by specific agonists such as 8-OH-DPAT facilitates ejaculation by blocking release of serotonin into the synapses (Ahlenius & Larsson, 1991).

This effect seems to be due to a decrease in serotonergic tone, since serotonin 1A receptors are found in the membrane of the cell body and dendrites of serotonergic neurons, indicating that they are presynaptic autoreceptors. Stimulation of presynaptic autoreceptors by serotonin results in a decrease in neuronal firing and consequent decrease in serotonin release, resulting in the same net effect as blocking the postsynaptic serotonin receptors. It should be noted, however, that not all serotonin 1A receptors are presynaptic; some of the effects of 8-OH-DPAT are mediated by postsynaptic receptors in the nucleus accumbens (Fernández-Guasti et al., 1992).

These findings indicate that serotonin acts as a brake for the production of ejaculation-orgasm by genital stimulation.

Is Premature Ejaculation Due to a Weak Serotonin Brake?

Normally, orgasm or ejaculation occurs only after a period of genital stimulation during copulation. This period varies greatly among species and among individuals within a species. Thus, rams and rabbits, which can be considered examples of "normally premature" ejaculators, ejaculate almost immediately (approximately one second) after intromission, while rats and men require considerable genital stimulation to overcome the tonic inhibitory effect, which is perhaps exerted by serotonin on the neural impulses originating from the genitals. It is tempting to speculate that rams and rabbits normally lack a genital-inhibitory serotonergic tone. Perhaps species differences in ejaculation latencies and amount of stimulation required to achieve ejaculation-orgasm are due to variations in balance between the tone of the serotonergic system and that of opposing systems, including the dopaminergic system, that establish an ejaculation threshold. Based on the same reasoning, perhaps men who experience premature ejaculation have a relatively low serotonergic tone, whereas those with anorgasmia have a relatively high serotonergic tone. A converse pattern of dopaminergic tone could have equivalent effects. It would be interesting to determine whether serotonin agonists would delay ejaculation in rams and rabbits, whether these animals' natural serotonergic tone is normally lower than that of rats and men, and whether the converse picture would be the case for their dopaminergic tone.

A Concise Neurological Model of Orgasm in Humans

Bearing in mind Alfred North Whitehead's recommendation to "seek simplicity but distrust it," we propose the following neurotransmitter model of orgasm in women and men. Admittedly simplified, it nevertheless provides a useful framework for understanding the effects of various drugs on orgasm. Abundant evidence points to dopamine as the key neurotransmitter involved in stimulating orgasm in humans. Thus, as mentioned above, administration of its precursor (L-dopa), agonists (apomorphine), dopamine releasers (amphetamine), or dopamine reuptake inhibitors (cocaine or bupropion) facilitates the expression of orgasm in men and women. Antipsychotics and many antidepressants that impair orgasm, on the other hand, possess the common property of blocking postsynaptic D2 or D4 receptors.

The neuroanatomical distribution of dopaminergic neurons in the brain and their projection sites are well known. The cell bodies of most dopaminergic neurons are located in the brainstem, particularly in the midbrain, and three fiber systems ascend from this structure to the forebrain. These fiber systems connect the ventral tegmental area in the upper midbrain (areas A10, A8) and substantia nigra (A9) with three forebrain areas: (1) the caudate nucleus and putamen (the neostriatum), (2) the limbic cortex (cingulate cortex, entorhinal cortex, and medial prefrontal cortex), and (3) a series of subcortical limbic structures (nucleus accumbens, amygdala, septum, and olfactory tubercle). Apparently, both D2 and D4 postsynaptic receptors are essential to orgasm, because drugs blocking either of these receptor subtypes inhibit orgasm. The cell bodies of the dopaminergic neurons that comprise the mesocortical and mesolimbic pathways originate in the ventral tegmental area. This brain region is activated during ejaculation in men, as measured by PET (Holstege et al., 2003), and its projection target region in the nucleus accumbens is activated during orgasm in women, as measured by fMRI (functional magnetic resonance imaging) (Komisaruk, Whipple, Crawford, et al., 2002, 2004). Both pharmacological and anatomical studies, therefore, strongly suggest that this dopaminergic system participates in the production of orgasm in humans. However, acute administration of drugs that increase dopaminergic activity (agonists, releasers, or uptake inhibitors) only occasionally induce orgasm in the absence of other factors. A possible exception is cocaine, which, when rapidly incorporated into the blood circulation, can induce the "cocaine rush" that some individuals report as feeling similar to genital orgasm.

Neurophysiological studies demonstrate that dopamine acts more as a neuromodulator of sensory input than as an excitatory neurotransmitter. The failure of dopaminergic drugs to produce orgasm by themselves is probably due to the fact that dopamine does not "turn on an orgasm switch." Rather, dopamine produces a sensitization to sensory stimuli (Antelman & Rowland, 1977). It opens and augments the flow of sensory impulses generated by genital stimulation and other sexually relevant stimuli to activate reward-pleasure limbic system circuits. Recent studies support this idea. Thus, Hagemann et al. (2003) found that in men with erectile dysfunction, the dopamine agonist apomorphine enhanced the response in the frontal cortex and the rostral anterior cingulate cortex (measured with PET) to the presentation of a sexually stimulating video. The authors concluded that apomorphine enhances the brain response to sexual stimuli.

Serotonin plays an essential role in the inhibitory regulation of orgasm. Thus, all antipsychotic and antidepressive drugs that enhance serotonergic activity, in most cases by blocking serotonin reuptake, tend to produce anorgasmia. Inhibition of orgasm is mediated by interaction of serotonin with serotonin 2 receptors. That this molecular process is critical for inhibition of orgasm is supported by the finding that agents such as cyproheptadine that block the action of serotonin on the serotonin 2 receptors almost immediately counteract the inhibitory effect of antidepressants on orgasm. The "exception that proves the rule" is found in the case of nefazodone, which, unlike the other SSRIs, does not inhibit orgasm. Nefazodone, in addition to blocking the reuptake of serotonin, also happens to block the serotonin 2 receptors, thereby counteracting the effect of the accumulated serotonin and preventing it from inhibiting orgasm.

We speculate further that endogenous serotonin normally exerts a tonic inhibitory effect on orgasm and that a transient decrease in serotonergic activity is necessary to enable the occurrence of orgasm. Indeed, buspirone, a serotonin 1A autoreceptor agonist (Zifa & Fillion, 1992), most likely exerts facilitatory effects on orgasm by decreasing the release of serotonin into the synapse.

Significant anatomical interactions exist between serotonergic and dopaminergic neurons. Serotonin exerts a tonic inhibitory effect on dopaminergic neurons at various sites. Some serotonergic fibers innervate dopaminergic neurons in the ventral midbrain. These dopaminergic neurons project their long axons forward into the striatum, limbic, and cortical areas (Jacobs & Azmitia, 1992). Serotonin input to these areas may inhibit dopaminergic neuronal firing.

Serotonin can also inhibit dopamine release via presynaptic serotonin receptors on dopaminergic neurons (Alex et al., 2005). This inhibitory process by presynaptic inhibition (through axoaxonic connections) has been well studied in the basal ganglia, and it is the basis of a serotonin-dopamine disorder theory proposed to explain Tourette syndrome and obsessive-compulsive disorders. Bodily tics or sudden obtrusive thoughts ("mental tics") observed in these disorders are thought to be due to transient predominance of the dopaminergic system over the serotonergic system. Extending this model to orgasm, when the balance between serotonin and dopamine is tilted toward serotonergic predominance, a tendency to delayed orgasm or anorgasmia may occur. By contrast, when dopaminergic uncontrolled activity prevails, premature ejaculation may occur. This could account for the ability to correct premature ejaculation either by increasing serotonergic tone or by blocking dopaminergic activity.

11

Effects of Medication

Kinsey and his colleagues stated in 1953 that sexual arousal and orgasm involve the entire nervous system and thus all parts of the body. Although this statement may seem exaggerated, recent studies of brain activity by functional magnetic resonance imaging (Komisaruk, Whipple, Crawford, et al., 2002, 2004) and positron emission tomography (Holstege et al., 2003) indicate that in both men and women, a large number of brain structures are activated, and others inactivated, during orgasm. This reveals the complexity of the neural circuits participating in orgasm and the many neurotransmitters and neuromodulators that could be involved. The complexity of the neural circuitry involved in orgasm makes its fine-tuning highly vulnerable to drugs that affect synaptic transmission. This may be the reason why most drugs impair, rather than improve, sexual response. In general, drug-related sexual disorders are caused by prescription drugs such as antipsychotics and antidepressants or by substances of abuse such as opiates, alcohol, and cocaine.

Since the 1950s, many medical reports have associated the use of some drugs (medicines, recreational drugs, herbs, etc.) with alterations in sexual response. Psychotropic drugs were among the first to be recognized as impairing sexual functioning, and we consider these first. But beyond their historical significance, they are important to discuss because these drugs, through their mode of action in producing anorgasmia, have aided more than other drugs in our understanding of the biochemical basis of orgasm in humans.

Neuroleptics and Atypical Antipsychotics

Neuroleptics (a broad category of antipsychotic drugs) were discovered by accident in the 1950s when Henri Laborit administered chlorpromazine, for its antihistamine effect, to patients being prepared for major surgery (Laborit & Huguenard, 1951). He observed that the chlorpromazine provoked mental effects such as decreased interest in the environment. This observation encouraged his colleague Jean Delay to test the action of this drug for antipsychotic potential (Delay, Deniker & Harl, 1952). They found chlorpromazine to be a highly effective antipsychotic. Its consequent widespread use created a revolution in the treatment of psychosis, particularly schizophrenia. Chlorpromazine was found to have its beneficial effects in suppressing the "positive symptoms" of schizophrenia (aggressive behavior, delusions, and hallucinations), not through its weak antihistamine action but by its antagonistic effect on dopamine postsynaptic receptors.

Chlorpromazine opened the door to the production of many other drugs with antipsychotic effects. (Table 1 lists the most widely used antipsychotics in Europe and the United States and their effect on sexual response.) Soon afterward, it was found that chlorpromazine—and, somewhat later, that other antipsychotics—had significant undesirable side effects on motor and muscular activity,

Table 1. Antipsychotic drugs and sexual disorder

Class	*Generic name*	*Trade name*	*Sexual disorder*
Phenothiazines	Chlorpromazine	Thorazine	++
	Fluphenazine	Prolixin	++
	Thioridazine	Mellaril	+++
Butyrophenone	Haloperidol	Haldol	+
Dibenzoxapine	Loxapine	Loxitane	++
New antipsychotics	Clozapine	Clozaril	0/+
	Olanzapine	Zyprexa	0/+
	Risperidone	Risperidal	+

0 no effect; + moderate effect; ++ strong effect; +++ intense effect

inducing parkinsonism (Parkinson's-like symptoms) and tardive dyskinesia (late-developing movement disorders). The muscle stiffness and uncoordination or lack of movement produced by these drugs was referred to as neurolepsis, and the term *neuroleptic* was coined by Laborit and Huguenard (1951) for the antipsychotic drugs that produce the undesirable motor effects. The effects were shown to be produced by the blockage of D2 receptors in the striatum (a subcortical region of the brain containing the putamen and caudate nucleus), which controls motor activity (D. Hartman, Monsma & Civelli, 1996; Jentsch & Roth, 2000). The striatum is a component of the extrapyramidal motor system (as distinct from the pyramidal, or corticospinal, motor system that originates in the motor cortex). The striatum receives dopaminergic fibers via the nigrostriatal pathway, which consists of axons originating in neurons located in the midbrain substantia nigra (Bjorklund & Lindvall, 1984).

Patients treated with antipsychotics frequently reported sexual disorders, though initially they were not taken seriously. Men frequently complained of erectile and ejaculatory problems. For example, in one study, half of the patients receiving thioridazine (Mellaril) reported disturbances in ejaculation (Kotin et al., 1976). Some of these sexual effects, such as erectile disorder, are not due to the effect of these drugs on the striatum or on dopamine receptors. Instead, they are due to the blockage of muscarinic cholinergic (acetylcholine) and alpha-adrenergic (norepinephrine) receptors in the peripheral nervous system or spinal cord that are related to erection and seminal emission. Anorgasmia and alterations in libido and sexual gratification were also found in men and women treated with neuroleptics (Ghadirian, Chouinard & Annable, 1982; Shen & Sata, 1990).

Adverse effects on orgasm have been observed with both low-potency neuroleptics (i.e., those having little effect on cholinergic receptors), such as chlorpromazine and thioridazine, and high-potency antipsychotics (having strong effects on cholinergic receptors), such as fluphenazine and haloperidol. Anorgasmia caused by antipsychotics is apparently due to their effect not on the striatum but rather on the limbic cortex, a separate region to which dopaminergic neurons project (Bjorklund & Lindvall, 1984). The cell bodies of dopaminergic neurons that project to the limbic cortex are located in a different part of the midbrain (not the substantia nigra), in a region termed the "ventral tegmental area." Their axons form the mesolimbic dopaminergic pathway.

Concern over the significant undesirable side effects of neuroleptics led to a search for antipsychotics that would selectively block dopaminergic activity in the mesolimbic ("antipsychosis") pathway but not in the nigrostriatal ("motor") pathway. Several compounds, known as "atypical" antipsychotics, were synthesized that did indeed have preference for the mesolimbic over the nigrostriatal pathway. Atypical antipsychotics are drugs that, at doses producing clear antipsychotic effects, lack neuroleptic (i.e., parkinsonian) effects. Particularly interesting is clozapine, an atypical antipsychotic that is highly effective in treating psychosis without undesirable motor effects (Stahl, 1999).

Clozapine (Clozaril) is a complicated drug that interacts with at least nine different receptors, and yet its antipsychotic effect seems to be due to its action on the dopaminergic system selectively. Clozapine has a much greater affinity for D4 receptors than for D2 receptors. The D4 receptor is similar to the D2 but has a different regional distribution in the nervous system. Thus, D4 receptors are found in the limbic cortex (mesofrontal and cingulate cortices) but to a much lesser extent in the striatum, a finding that accounts for clozapine's lack of parkinsonian effects (Pilowsky et al., 1997). This drug blocks the dopamine effect in the limbic system, thereby dampening the positive symptoms of schizophrenia. However, clozapine also induces orgasmic disorder in women and men.

These observations, taken together, led to the conclusion that limbic D4 receptors are involved in orgasm. The orgasmic disorder produced by clozapine is much weaker than that observed with classical antipsychotics, at least in men. Thus, Aizenberg, Modai, et al. (2001) found that male patients with schizophrenia experienced less severe orgasmic disorder (measured as number of orgasms per month and frequency of orgasm during sexual intercourse) than comparable patients treated with the classical antipsychotics. Similarly, in a group of male patients, Wirshing et al. (2002) reported significantly fewer cases of orgasmic disorder in those treated with clozapine than in those treated with haloperidol or risperidone.

Use of Antipsychotics in Treating Paraphilias

Both patients and clinicians perceive the impairment of sexual response that is produced by antipsychotic drugs as an unwelcome side effect. Yet this effect can be used advantageously to ameliorate certain types of sexual problems. The fact that antipsychotic drugs decrease sexual libido and delay or block orgasm prompted some clinicians to use these drugs to treat "hypersexual" individuals. Thus, it has been reported that antipsychotics control the behavior of male subjects with various paraphilias. (According to the DSM-IV-TR [2000], paraphilias are mental disorders "characterized by sexual fantasies, urges or behaviors involving non-human objects [coprophilia, Fetishism, Transvestic Fetishism], suffering or humiliation [Sexual Sadism, Masochism], children [Pedophilia], or other non-consenting person [Voyeurism, Frotteurism, Exhibitionism].") Some antipsychotics have also been used successfully in the treatment of premature ejaculation (PE), a condition in which the ejaculation-orgasm threshold is abnormally low.

Mechanism of Action of Antipsychotics on Orgasm

Antipsychotics not only interfere with elicitation of orgasm but impair erection and ejaculation by blocking spinal and peripheral cholinergic and adrenergic receptors (Stahl, 1999). This effect is in addition to their interference with the perceptual experience of orgasm by their action in the brain. The effects of drugs on orgasm provide clues to the biochemical mechanisms involved in the production of this unique mental state.

The common action of all antipsychotics in producing anorgasmia is their blockage of dopamine receptor subtypes: D2 (blocked by typical antipsychotics) or D4 (blocked by atypical antipsychotics). While blockage of D2 and D4 receptors in the limbic cortex (which includes the medial prefrontal cortex and the cingulate cortex) is related to the disappearance of the positive symptoms of schizophrenia (hallucinations, delusions), it also interferes with the normal production of orgasm. The relationship between orgasm and hallucinations suggests that both processes involve the activation of common neurons that have dopamine receptors and receive dopaminergic input. This confounding of effects makes the task of designing pharmaceutical agents that can differentially affect these processes difficult.

Thus, anorgasmia does not seem to be related to the neuroleptic motor effects induced by typical antipsychotics acting on the nigrostriatal dopaminergic system (in which neurons in the substantia nigra send axons to the striatum). Furthermore, the atypical antipsychotics—those *lacking* extrapyramidal motor effects—such as clozapine are less disruptive to orgasm than the typical antipsychotics (Aizenberg et al., 2001; Wirshing et al., 2002).

While the role of the dopaminergic mesocortical and mesolimbic systems in the expression of orgasm seems well founded, several puzzling observations remain to be explained. For example, although blockage of dopamine receptors occurs within hours of taking an antipsychotic, anorgasmia does not occur until days later. A comparably delayed action of the drugs was also observed for the improvement of the positive symptoms of schizophrenia.

This delayed action of antipsychotics led some investigators to propose the existence of a secondary reorganization of dopamine receptors in response to the initial effect. Reorganization of dopamine receptors in the membrane of receptive neurons may occur as a physiological adjustment to fluctuations in dopamine released at the synapse, which then affects the intensity of "stimulation" produced by the neurotransmitter. Thus, when an excessive concentration of dopamine persists in the synapse, it produces a compensatory *decrease* in the number of receptors, a process termed "down-regulation." Conversely, when dopamine is depleted or its action on the receptors is prevented by antagonists (such as antipsychotics), the consequence is a compensatory *increase* in the number of receptors, or "up-regulation." An up-regulation of dopamine receptors occurs following a persistent blockade of dopamine action by antipsychotics, and probably plays a role in the action of these drugs in both psychosis and sexual response.

However, recent findings suggest that the main event in the antipsychotic and sexual response effects of these drugs is not the up- or down-regulation of the receptors but rather blockade of the D2 or D4 receptors (Kapur & Mamo, 2003). A meta-analysis of several well-controlled studies revealed a much more rapid onset of action of both typical and atypical antipsychotics in improving psychotic symptoms. Indeed, well-controlled, double-blind placebo studies comparing the effects of haloperidol, olanzapine, and placebo showed a clear effect of the antipsychotics within as little as twenty-four hours. This indicates that blocking dopaminergic activity or D2 or D4 receptors produces an almost immediate antipsychotic effect (Kapur, 2004). It seems likely that a careful study with an adequate questionnaire would also reveal a comparably early effect of the blocking of dopaminergic activity on orgasm or sexual functioning.

Regardless of whether the action of antipsychotics on orgasm is based on a primary (receptor-blocking) or secondary (receptor up- or down-regulation) effect, there is little doubt that dopamine is a significant, if not essential, neurotransmitter for the production of orgasm.

An alternative mechanism that could account for the anorgasmia and sexual disorder produced by antipsychotics is the increased secretion of the anterior pituitary hormone prolactin into the bloodstream, which occurs in response to typical and some atypical antipsychotics (e.g., risperidone). High plasma levels of prolactin are associated with an inhibited sexual response in both sexes. However, the finding that clozapine does not increase prolactin levels (Melkersson, 2005) but nevertheless induces orgasmic disorder indicates that the process that increases prolactin is not essential to producing this disorder.

Problematic Antidepressants

A high proportion of individuals with depression have sexual disorders (Segraves, 1992, 1993). Indeed, in the nineteenth century, the French psychiatrist Pierre Janet considered impairment of sexual functioning to be a symptom of depression. The mechanism involved in sexual disorder during depression is not clear and could be considered part of the anhedonia (the failure to enjoy positive experiences) characteristic of this disease. If depression is related to a sexual disorder, effective treatment of the depression would be expected to improve sexual life.

In the 1950s, the first effective antidepressants became available in clinical practice. The initial information on the effect of antidepressants on sexuality derived from single case or anecdotal reports; nevertheless, subsequent studies supported these reports. Rather than correcting orgasmic disorders, antidepressants exacerbated the problem. Harrison et al. (1986) made the first well-controlled systematic study on the effect of antidepressants on the sexual response of women and men with depression. They compared the effect of a placebo with the tricyclic antidepressant imipramine (Tofranil) and the nonreversible monoamine oxidase (MAO) inhibitor phenelzine (Nardil). Their study used a well-designed questionnaire for evaluating sexual functioning. The incidence of patients reporting a significant decrease in sexual functioning was 30 percent with imipramine, 40 percent with phenelzine, and only 6 percent with the placebo. One of the main problems reported was the inability to experience orgasm by either intercourse or masturbation. An extensive series of reports and studies has confirmed these findings (e.g., Baier & Philipp, 1994; Lane, 1997), although the percentages of patients with sexual disorders in response to antidepressants varied widely depending on the methodologies used.

Antidepressants can be classified into three families, depending on both the cellular mechanisms involved in their action and their chemical structure: (1) inhibitors of MAO, (2) tricyclic antidepressants, and (3) selective serotonin reuptake inhibitors (SSRIs). We discuss them separately because their pharmacological profiles differ, as do their effects on sexual response. (Table 2 lists the antidepressants used in the United States and Europe and their effect on human sexual response.)

Monoamine Oxidase Inhibitors

Monoamines are chemicals that contain one amine group (meaning an "ammonia-containing" group, consisting of nitrogen and hydrogen). Monoamines are enzymatically broken down and

Table 2. Antidepressant drugs and sexual disorder

Class	*Generic name*	*Trade name*	*Sexual disorder*
Tricyclic	Amitriptyline	Elavil	++
	Clomipramine	Anafranil	++
	Desipramine	Norpramin	++
	Imipramine	Tofranil	++
	Nortriptyline	Pamelor	++
MAOI	Moclobemide	Aurorex	0
	Phenelzine	Nardil	+
SSRI	Citalopram	Celexa	+++
	Fluoxetine	Prozac	++
	Fluvoxamine	Luvox	++
	Paroxetine	Paxil	+++
	Sertraline	Zoloft	++
Atypical	Bupropion	Wellbutrin	0
	Mirtazapine	Remeron	0
	Nefazodone	Serzone	0
	Reboxetine	Edronax	0
	Trazodone	Desyrel	0
	Venlafaxine	Effexor	++
Antagonist	Buspirone	BuSpar	0

0 no effect; + moderate effect; ++ strong effect; +++ intense effect

thereby inactivated by the enzyme monoamine oxidase. MAO inhibitors block the degrading action of the enzyme. In the synapse, MAO inhibitors act by preventing the enzymatic breakdown of the monoamines—norepinephrine, dopamine, and serotonin—and thus increasing the availability, and consequently prolonging the effect of, these neurotransmitters at the postsynaptic neurons. MAO inhibitors were the first antidepressants to be introduced into clinical practice. The discovery of their antidepressant effect was serendipitous—they were initially tested as antituberculosis agents.

All the original MAO inhibitors, such as phenelzine (Nardil), bind to MAO irreversibly, thus permanently blocking its enzymatic function. An unfortunate result of this action is that the drug allows norepinephrine levels to increase at all receptor sites, including those where norepinephrine stimulates hypertension—a dangerous complication, particularly if a person taking the drug eats foods, such as cheese, that are rich in tyramine, a precursor of norepinephrine. In a substantial proportion of women and men, irreversible MAO inhibitors also produce significant undesirable sexual side effects, including anorgasmia.

A great advance in treatment for depression was made when *reversible* inhibitors of MAO, such as moclobemide (Aurorex), became available. Reversible inhibitors lack the troublesome hypertensive action yet retain their antidepressant effect. Also, moclobemide produces sexual disorders in only a small proportion of patients (fewer than 5%), making it an ideal drug for patients who have experienced troubling sexual disorders, such as anorgasmia, with other MAO inhibitors (Philipp, Kohnen & Benkert, 1993; Montejo et al., 2001). An unexpected side effect of moclobemide is the production of hypersexuality (Lauerma, 1995) and hyperorgasmia in a limited number of women and men.

Most MAO inhibitors increase the levels in the brain of all monoamines. Yet, norepinephrine and dopamine are generally considered to facilitate sexual response. Therefore, it seems likely that the orgasmic disorder associated with many MAO inhibitors is due to an increase in serotonergic activity that eclipses the otherwise facilitatory effects of elevated norepinephrine and dopamine. Indeed, evidence from both animal and human studies indicates that serotonin alone, acting through serotonin 2A receptors, impairs normal levels of sexual functioning.

Tricyclic Antidepressants

Tricyclic antidepressants (TCAs) were initially tested as neuroleptics (i.e., antidopaminergics) for the treatment of schizophrenia—with poor results. But they were found to be good antidepressants. Following the finding of Harrison et al. (1986) that imipramine

(Tofranil) induced sexual disorders, other TCAs have also been reported to produce orgasmic disorder (delayed orgasm and/or anorgasmia). Amitriptyline, trimipramine, clomipramine, and desipramine have been reported to induce anorgasmia in women and men (J. M. Ferguson, 2001). Moreover, due to the pharmacological profile of these drugs, they affect most aspects of sexual functioning.

Tricyclic antidepressants modify several neurotransmitter systems, both in the peripheral nervous system (i.e., the autonomic and other peripheral nerves) and in the central nervous system. They inhibit the reuptake of monoamines and also block muscarinic cholinergic and histaminergic (histamine) receptors (Stahl, 1999). It seems likely that their anticholinergic effect produces the interference with erection and ejaculation often reported by men who are taking these drugs (Labbate, Croft & Oleshansky, 2003). The corpora cavernosa of the penis are innervated by cholinergic nerves and contain cholinergic muscarinic receptors that can be blocked by TCAs. This accounts for the observation that the administration of bethanecol, a cholinergic stimulatory agent, reverses the erectile and ejaculation difficulties produced by some TCAs (e.g., imipramine). On the other hand, nearly all TCAs enhance serotonergic activity, and therefore it is probable that they also impair the cognitive experience of orgasm by stimulating serotonin 2A receptors.

Selective Serotonin Reuptake Inhibitors

Some years ago it was believed that depression was due to a deficiency in catecholamines, particularly norepinephrine and serotonin. This belief was based on the fact that the first successful antidepressants—MAO inhibitors and TCAs—increased the concentrations of these neurotransmitters. Although there is little doubt that these neurotransmitters, particularly serotonin, are involved in depression, it now seems that the problem lies with their receptors rather than with the neurotransmitters themselves

(Stahl, 1999). Antidepressants boost the levels of monoamines almost immediately, but their effect on mood has a significant delay, usually two or three weeks after blood levels of monoamines have increased. It is thought that depletion of norepinephrine and serotonin produces an up-regulation (i.e., an increase in the number) of postsynaptic receptors, which in some way is related to depression. The serotonin 2A receptor is implicated in this process, on the basis of evidence that high concentrations of this receptor are found in the frontal cortex of people who have committed suicide (Arango, Underwood & Mann, 1992). This finding could be interpreted as the result of a low level of serotonergic activity, which would produce a compensatory increase in the production of serotonin receptors (i.e., up-regulation). Increasing the levels of serotonin by blocking its enzymatic breakdown, or by blocking its reuptake, will produce a decrease (down-regulation) of these receptors and reverse the depression.

Because the original antidepressants (MAO inhibitors and TCAs) produced troublesome side effects, research was concentrated on developing drugs that would selectively boost serotonin concentration. These drugs were called selective serotonin reuptake inhibitors (SSRIs). Since the introduction of fluoxetine (Prozac), SSRIs have become the most highly used antidepressants, particularly for long-term use. Yet, as has gradually become evident, most SSRIs cause delayed orgasm or anorgasmia in a high proportion of patients (Preskorn, 1995; Rosen, Lane & Menza, 1999; J. M. Ferguson, 2001; Montejo et al., 2001), with minor or no effects on libido or sexual arousal (i.e., on penile erection or on vasomotor genital changes in women).

Mechanism of Action of Antidepressants on Orgasm

Many of the drugs that produce sexual disorders increase serotonin levels. The mechanisms by which they do this vary. MAO inhibitors increase the levels of serotonin by preventing its enzymatic destruction, while both TCAs and SSRIs increase the levels of serotonin by blocking its reuptake. There is evidence suggesting that serotonin 2A receptors are involved in sexual disorders, particularly anorgasmia. Serotonin 2A receptors modulate the neural firing of the postsynaptic neurons on which they are situated by the production of second messengers (J. R. Cooper, Bloom & Roth, 2003). Blocking these receptors with cyproheptadine, an antihistamine and potent serotonin type 2 antagonist, reduces or reverses the sexual disorders and anorgasmia produced by antidepressants (Arnott & Nutt, 1994; Woodrum & Brown, 1998).

The role of serotonin 2A receptors in inhibiting sexual response, including orgasm, is also supported by the action of a class of agents known as phenylpiperazines, which are potent blockers of the serotonin 2A receptors. These drugs also block serotonin reuptake, although less effectively than TCAs or SSRIs, and are thus termed serotonin 2 antagonists/reuptake inhibitors (SARIs). When serotonin reuptake is inhibited, the boost of available serotonin results in the stimulation of all serotonin receptors, some of which, such as serotonin 1A receptors or serotonin 2D receptors, have been implicated in the *facilitation* of sexual response. This may account for the observation that trazodone (the first member of the SARI group to be discovered) occasionally can produce erections, even priapism (prolonged erections with soreness and pain) (Warner, Peabody & Whiteford, 1987). A promising new SARI is nefazodone which, like trazodone, is a powerful antagonist of the serotonin 2A receptor. As might be expected from its pharmacological profile, nefazodone is one of the few antidepressants that do not produce significant anorgasmia (Montejo et al., 2001; Baldwin, 2004). On the contrary, nefazodone stimulates spontaneous ejaculation in some patients.

Sexually Beneficial or Benign Antidepressants

Premature ejaculation is one of the most common sexual complaints. According to a study by A. J. Cooper, Cernovsky, and Colussi (1993), PE affects a significant proportion of men. Since the antidepressants act by increasing serotonergic activity, their inhibitory effect on orgasm reveals the existence of a natural serotonergic orgasm-inhibitory mechanism. The existence of this inhibitory mechanism is also supported by the finding that several antidepressants can prevent PE. Thus, the SSRIs fluoxetine, paroxetine, and sertraline are effective in increasing the latency of ejaculatory orgasm from less than one minute, which is the approximate latency for a diagnosis of PE, to between two and six minutes (McMahon, 1998; McMahon & Touma, 1999a, 1999b). Delay of orgasm by antidepressants occurs not only in men with PE but also in men who ejaculate with typical, longer latencies after the initiation of sexual intercourse. Thus, Waldinger, Zwinderman, and Olivier (2003) found that paroxetine increased ejaculation latency by 420 percent in men with PE and by 480 percent in men with typical ejaculation.

The site of action of serotonin on serotonin 2A receptors to delay or suppress orgasm is not known. Serotonin 2A receptors have a wide distribution in the central nervous system, from the spinal cord to the cerebral cortex. Occupation of serotonin 2A receptors by the neurotransmitter stimulates neuronal activity (J. R. Cooper, Bloom & Roth, 2003). This suggests that inhibition of these neurons is necessary to trigger orgasm. Indeed the SARI nefazodone, which blocks serotonin 2A receptors in addition to blocking serotonin reuptake, reverses the sertraline-induced anorgasmia when administered one hour before sexual activity. This is probably due to occupation of the serotonin 2A receptors by the nefazodone, which blocks the subsequent inhibitory effect of the sertraline.

Moclobemide. Fortunately, not all effective antidepressants produce sexual disorders. In each class of antidepressants, some drugs are almost free of sexual side effects, or even have a facilitatory effect on sexual response. Thus, reversible inhibitors of MAO type A, such as moclobemide, do not impair sexual response (Montejo et al., 2001). On the contrary, some reports (e.g., Lauerma, 1995) describe facilitatory effects on all components of sexual response, and "hyperorgasmia" has been described in a woman taking moclobemide.

Reboxetine. The mechanism of action of this antidepressant is inhibition of norepinephrine reuptake, thereby increasing its levels in the synapse, without affecting serotonin. Reboxetine does not seem to produce sexual disorders (A. H. Clayton et al., 2003).

Bupropion. This well-established antidepressant is not known to produce sexual disorders. Like cocaine, it blocks the dopamine reuptake mechanism, thereby increasing the level of dopamine in the synapse. Well-controlled, double-blind studies, with or without placebo, have shown that even prolonged treatment with bupropion does not affect orgasm (Segraves et al., 2004). Moreover, bupropion is an effective antidote for the sexual disorders induced by most SSRIs (Ashton & Rosen, 1998; A. H. Clayton et al., 2003). Bupropion, like other antidepressants (TCAs and MAO inhibitors), stimulates the dopaminergic system by selectively blocking the reuptake of dopamine without affecting serotonergic activity.

Nefazodone. By contrast with some other antidepressants, nefazodone and other SARIs block the reuptake of serotonin, thus increasing serotonin levels in the synapse but also shielding the serotonin 2A receptors from the action of that increased serotonin. A large series of well-controlled studies indicated that sexual side effects are minimal or nonexistent with this drug (Preskorn, 1995; Feiger et al., 1996; Montejo et al., 2001). A side effect of nefazodone is that on occasion it induces spontaneous ejaculation.

Buspirone. This interesting drug was used as an anxiolytic (anxiety-relieving drug), with distinct antidepressant effects. It acts as an agonist of the serotonin 1A receptors, which have been related to sexual stimulation in animal models. Indeed, in well-controlled studies, buspirone was effective in counteracting the negative side effects on sexual response of other antidepressants (Landen et al., 1999).

St. John's Wort. This herb is evidently effective in treating minor depression (H. L. Kim, Strelzer & Gaebert, 1999). It contains the active ingredient hyperforin, which increases serotonin levels. It does not seem to produce overt effects on sexual functioning (Linde et al., 1996; Muller et al., 1998).

12

Counteracting Medication Side Effects

Many antipsychotic and, particularly, antidepressant drugs have adverse effects on sexual response and orgasm. It is often possible to select medications that have little or no such side effects. Or, if patients and their doctors find that a drug with unwanted sexual side effects is the most suitable because of its other, desirable therapeutic effects, several strategies have been proposed to deal with the accompanying sexual disorders (D. O. Clayton & Shen, 1998).

A physician may first recommend continuing with the treatment for a protracted period because, in many cases, the patient may eventually tolerate the drug or experience spontaneous remission of the drug-related sexual disorder. This recommendation is sound for patients taking monoamine oxidase (MAO) inhibitors or selective serotonin reuptake inhibitors (SSRIs), but not when using tricyclic antidepressants (TCAs). The anorgasmia produced by TCAs does not remit spontaneously.

A different strategy is to reduce the medication dose in the hope of maintaining therapeutic efficacy while reducing the deleterious effects on sexual functioning. Another strategy is for the patient, when sexual activity can be planned, to take a "drug holiday," discontinuing treatment two to three days before a sexual interaction. Obviously, this strategy does not work for individuals having frequent sexual activity. "Drug holidays" may be particularly effective with antidepressants that have a relatively short half-life, such as sertraline, venlafaxine, and fluvoxamine. (The half-life is the time it takes, after discontinuing treatment, for the level of drug in the bloodstream to reach half its original concentration, or for the effect of the drug on some measurable quantity to diminish by half.) Thus, in one study, stopping sertraline for at least 48 hours reversed orgasm disorder and allowed sexual satisfaction in 80 percent of patients who previously suffered adverse sexual side effects. This strategy, however, would be inappropriate for a drug such as fluoxetine, which has a longer half-life of 48 to 72 hours (Shen, 1997).

In some cases, desirable outcomes are obtained by switching from one antidepressant to another, usually one of the same class, as has been reported for TCAs—for example, switching from imipramine to desipramine (Sovner, 1983). A more logical approach is to change from the antidepressant that is producing orgasmic disorder to one that is known to lack sexual side effects, such as nefazodone, mirtazapine, or bupropion. This procedure has been effective in some cases. Several studies indicate a marked improvement in sexual activity (including orgasm) when nefazodone or mirtazapine is used to replace another SSRI (Labbate, Croft & Oleshansky, 2003; Baldwin, 2004).

The decision to switch antidepressants is difficult when a patient is experiencing a good antidepressive effect but has an undesirable sexual response. Numerous studies and clinical trials have been performed in search of drugs that could counteract the orgasmic disorder produced by the antidepressants, but few well-controlled studies have investigated such "pharmacological antidotes." Nevertheless, the efficacy of some of these "antidotes" is validated by agreement among case reports.

When the antidepressant interferes with penile erections, sildenafil (taken 30 minutes before sexual intercourse) has been reported to improve erectile function (Ashton & Bennett, 1999; Marks et al., 1999). Sildenafil is an inhibitor of the enzyme phosphodiesterase-5, which normally inactivates the second messenger cGMP. The production of cGMP is stimulated by nitric oxide, a neurotransmitter that is released onto the smooth muscle of the penis. Normally, cGMP relaxes the smooth muscle that surrounds penile arterioles, thereby increasing the blood flow to the penis and stimulating erection. By inhibiting phosphodiesterase-5, sildenafil blocks the breakdown of cGMP and consequently augments its action, relaxing smooth muscle and thus promoting erection. This effect might be indirect, resulting from the psychological benefit obtained from its desirable peripheral effect of intensifying erection.

Another drug, yohimbine, counteracts sexual disorder by acting on both the peripheral and the central nervous system. Initially, this alpha-2-adrenergic antagonist was used to treat erectile dysfunction. Subsequently, it was found to improve arousal and orgasm (Hollander & McCarley, 1992; Jacobsen, 1992). Yohimbine acts by blocking the autoreceptor for norepinephrine, which *dis*inhibits the release of norepinephrine from the neuron, thereby increasing its level in the synapse. This has the effect of facilitating orgasm in women and men. Yohimbine functions in this way when given two to four hours before sexual activity.

As there is good evidence supporting a role of serotonin 2A receptors in delay of orgasm and in anorgasmia, the effect of agents that block these receptors has been tested. Thus, nefazodone, mianserin, and mirtazapine, with antidepressant effects by themselves, relieve orgasmic disorder caused by concurrent treatment with SSRIs or the antidepressant venlafaxine (Aizenberg, Gur, et al., 1997; Reynolds, 1997). Reversal of the anorgasmia induced by the antidepressant sertraline has also been reported with the antidepressant nefazodone, taken one hour before sexual intercourse (Reynolds, 1997). Moreover, cyproheptadine, an antihistamine with powerful serotonin 2A antagonism, restores normal orgasm and reverses other sexual side effects caused by MAO inhibitors, TCAs, and SSRIs, when taken before sexual activity (Arnott & Nutt, 1994; Woodrum & Brown, 1998).

It is evidently the serotonin 1A receptor that mediates the *stimulatory* effect of serotonin on sexual response. This property makes the serotonin 1A receptor a possible target for drugs aimed at correcting medication-induced sexual disorders. One of these drugs is buspirone, the anxiolytic that acts as an agonist at the serotonin 1A receptor. Buspirone has been found to reverse the negative effects of antidepressants on both libido and orgasm (Norden, 1994). Although buspirone also affects other neurotransmitter systems, its main effect seems to be mediated through the serotonergic system.

The other well-established mechanism related to the anorgasmia produced by antipsychotics and antidepressants is inhibition of, or a decrease in, dopaminergic action. Postsynaptic D2 receptors in the limbic system (prefrontal cortex, cingulate gyrus, and nucleus accumbens) participate in the production of orgasm.

Bupropion is an antidepressant that lacks serotonergic activity and does not induce sexual disorders (Montejo et al., 2001). On the contrary, some case reports suggest that it could be stimulatory for orgasm (Segraves et al., 2004) and ejaculation. Bupropion, particularly its hydroxylated metabolite, is a reuptake inhibitor of both noradrenaline and dopamine, and this effect is responsible for its antidepressant action. Its dopaminergic activity is most likely responsible for its stimulatory effect on sexual response. Bupropion, chronically administered, is effective in improving the sexual phases of desire, arousal, and orgasm (for reviews, see J. M. Ferguson, 2001; A. H. Clayton et al., 2003). Other dopaminergic agents, such as amantadine, have also been reported to counteract the undesirable sexual side effects of the antidepressants.

13

Recreational Stimulant Drugs and Orgasm

The extensive and persistent use (and abuse) of "recreational" drugs for enhancing sexual response demonstrates the widespread belief that these substances are effective for this purpose. Researchers have studied the effects of many of these drugs on orgasm, and the results shed light not only on the experiences for drug users but also on the nature of orgasm. Here, we review the major recreational drugs and what is known of their effects on orgasm.

Marijuana

Marijuana is obtained from the plant *Cannabis sativa*. Marijuana and alcohol (a depressant) are the most widely used recreational drugs. A mild hallucinogen in the doses usually consumed, marijuana is smoked to deliver its active components to the brain. The most important cannabinoid in marijuana is delta-9-tetrahydrocannabinol (THC). In mild dosages it tends to produce a sense of well-being, a loosening of social tensions, and a feeling of achieving special insight. Its acute effect on sexual response varies considerably depending on the smoker's personality and experience, the social context in which the drug is taken, and the prior expectations about its effect. A substantial proportion of persons claim that marijuana enhances and enriches their sexual experience. Marijuana users report more intense, prolonged, and rewarding orgasms than those experienced without the drug. Users report an increased awareness of peripheral activity such as muscular contractions in the genital area during orgasm (Kaplan, 1974). Other users report that marijuana delays orgasm and that when it occurs, it feels highly pleasurable.

By contrast with these mostly anecdotal reports, there are research data indicating that chronic users of marijuana often have orgasmic disorders. In a recent analysis of sexual disorders resulting from drug use in a sample of 3,004 adult men and women, neither marijuana nor alcohol decreased libido, but their use was clearly associated with anorgasmia (Johnson, Phelps & Cottler, 2004). This effect on sexual activity may be the direct result of the "amotivational syndrome" described by frequent users and characterized by a generally reduced drive, a condition that most likely impairs daily life activity, including sexual activity.

Ecstasy (MDMA)

Known by the common name Ecstasy, MDMA (3,4-methylenedioxymethamphetamine) is a derivative of amphetamines, but it produces subjective and physiological effects different from those of the amphetamines. It exerts potent effects on social and sexual response. According to recreational users, it tends to increase sociability and generates a positive mood, effects that favor sexual interactions. Some recent studies (e.g., Zemishlany, Aizenberg & Weizman, 2001) designed to test Ecstasy's reputed aphrodisiac effect found that libido was moderately to greatly increased in more than 90 percent of women and men using the drug. However, penile erection was impaired in 40 percent of the men. Orgasm was usually delayed but perceived as more intense and of a richer quality than without the drug. The mechanisms through which Ecstasy produces these effects are not well understood.

Poppers (Amyl Nitrite)

Amyl nitrite is a highly volatile liquid that is used medically to relieve the discomfort or pain of angina pectoris (cardiac-related chest pain). This effect is achieved by its rapid, potent vasodilator action. Medicinal preparations are sold in small, single-use glass ampules. In order to inhale the contents, the user breaks open the ampule, producing a slight popping sound. Users of amyl nitrite claim that when it is "popped" at the beginning of orgasm it increases orgasmic intensity. The main mechanism underlying this effect is vasodilation of the genital area, a process that can intensify erection and thereby enhance sensation. The precipitous and marked drop in blood pressure produced by the inhalation of amyl nitrite probably also produces a sudden and direct arousing effect in the brain, which in the sexual context may intensify perceptions. People using amyl nitrite must be warned not to use it with sildenafil, because of the synergistic hypotensive (blood pressure–lowering) effect.

Amphetamines and Cocaine

Psychostimulants, particularly the amphetamines, were widely used in medical practice until the late twentieth century. At present they are used medically only as anorexiants (for weight reduction), and some derivatives, such as dextroamphetamine (Dexedrine), are used for treatment of attention-deficit hyperactivity syndrome and narcolepsy.

Psychostimulants are widely used as recreational drugs. Cocaine is a potent stimulant that increases stamina (an effect for which Sigmund Freud used himself as a research subject) (Chiriac, 2004). Cocaine is described as producing a sense of mental acuity and a state of euphoria—qualities that are highly rewarding and the basis for its abuse. Amphetamines have mental effects similar to those of cocaine, although the state of euphoria is less intense. Both cocaine and amphetamines act primarily by increasing dopaminergic activity; cocaine inhibits dopamine reuptake, whereas amphetamine initially stimulates release of the neurotransmitter and secondarily inhibits its reuptake. Acute intravenous administration of cocaine induces a highly pleasurable feeling ("rush") that some individuals equate with sexual orgasm. Both cocaine and amphetamine administered before sexual intercourse are reported to facilitate orgasm and enhance the quality of this experience (N. S. Miller & Gold, 1988; Kall, 1992). Kall (1992) studied 29 young men who used amphetamine more than once a week for at least six months. Twenty-seven of these individuals experienced sexual activity while using amphetamine; of these, 21 reported intensified orgasms and 23 reported that the drug prolonged intercourse. Chronic use of cocaine or amphetamines, however, induces sexual disorders and anorgasmia in a high proportion of addicts. It often produces symptoms similar to paranoid schizophrenia.

Nicotine

The effects of cigarette smoking or nicotine itself on sexual response and orgasm have not been systematically researched, probably because nicotine does not seem to have a prominent effect. This lack of effect is surprising considering nicotine's potent ability to mimic the action of the neurotransmitter acetylcholine. An early study of women by Fischer (1973) failed to find a clear effect of smoking on most of the parameters of sexual response that were analyzed. A curious exception was a significant positive correlation between the number of cigarettes smoked and women's reported intensity of orgasm, possibly involving a rewarding effect of dopaminergic stimulation.

14

Depressant Drugs and Orgasm

A large variety of drugs that produce *depression* of brain activity are used for medical purposes or recreationally. Practically all of the depressant drugs influence human sexual response, although some discrepancy exists as to the nature of their effects, particularly depending on the dose, the frequency with which they are taken, and the social context in which they are used.

Depressant drugs can be categorized into two families: one interacting with the GABA-A receptor (stimulating an increase in GABAergic activity) and the other acting on opiate receptors that are normally activated by endogenous opioid neuropeptides such as enkephalins, endorphins, or dynorphin. GABA (an acronym for gamma-aminobutyric acid) is perhaps the most prevalent *inhibitory* neurotransmitter in the nervous system. It has been estimated that GABA is present in 30 to 40 percent of the synapses in the central nervous system (Best, 1990). An inhibitory neurotransmitter reduces, rather than stimulates, the activity of the neurons with which it interacts.

Alcohol

Alcohol is a drug with extremely complicated effects. Surprisingly, although alcohol was probably the first psychotropic used by humans and is the most widely used and abused recreational drug, its pharmacology is still poorly understood. Undoubtedly it affects many neurotransmitter systems (including receptors and enzymes) and has nonspecific effects at the neuronal membrane level. But some data suggest that alcohol acts as a type of modulator of the GABA-A receptor, thereby enhancing the inhibitory activity of this neurotransmitter. In addition, alcohol reduces the effect of a different, *excitatory* neurotransmitter—glutamate—which acts on another type of receptor: the NMDA (*N*-methyl-D-aspartate) receptor. Therefore, alcohol not only increases neuronal inhibition through the GABA-A receptor but also reduces excitation through the NMDA receptor—a dual depressive action.

Folklore and popular belief have ascribed to alcohol an aphrodisiac effect. However, alcohol is a potent central nervous system depressant, which when ingested before a sexual interaction, even in moderate amounts, interferes with penile erection (Wilson, 1981). Its reputation as an aphrodisiac is based on the fact that it initially reduces anxiety and may disrupt sociocultural inhibitions, thus disinhibiting, releasing, and transiently increasing libido. As the effect of alcohol progresses, however, the imbibing individual tends to become incapacitated for sexual performance (D. E. Smith, Wasson & Apter-Marsh, 1984; N. S. Miller & Gold, 1988). Its effects on sexual response are well summarized by Shakespeare: "it provokes the desire, but it takes away the performance."

A recent study on a large epidemiological sample revealed that in men, when alcohol users were able to engage in sexual intercourse, orgasm was delayed or inhibited (Johnson, Phelps & Cottler,

2004). In women, daily alcohol intake was directly (positively) correlated with the amount of genital stimulation required to experience orgasm (Fischer, 1973). A different study found that in moderate amounts, alcohol can facilitate all phases of sexual response in women, including orgasm; however, alcohol only increased the women's *subjective* level of sexual arousal (Malatesta et al., 1982).

It seems reasonable that alcohol ingestion before sexual intercourse has been recommended as a treatment for premature ejaculation. However, to our knowledge, no well-designed study to validate the use of alcohol to treat this problem has been reported.

Tranquilizers (Anxiolytics)

By contrast with alcohol, the mechanism of action of the benzodiazepine tranquilizers is well established. These drugs act as positive modulators of GABAergic activity by binding to a specific (benzodiazepine) site on the GABA-A receptor. This interaction increases the passage of chloride (Cl^-) ions through a chloride channel, which "hyperpolarizes" and thus inhibits the receptive neurons. This inhibition is responsible for the anxiolytic, anticonvulsant, and sedative effects of the benzodiazepines. These effects are related to a state of tranquility that leads some persons to abuse these drugs. As could be expected from their anxiolytic effect, low doses of benzodiazepines may facilitate social and sexual interactions in anxious individuals, even increasing their libido (Fava & Borofsky, 1991). Some studies (e.g., Matsuhashi et al., 1984) found that benzodiazepam ameliorated psychogenic erectile dysfunction, improving not only erectile function but also sexual satisfaction. This disinhibitory effect exerted by benzodiazepines has been reported in both men and women and may have facilitated sexual assaults ("date rape") under the effect of benzodiazepines: "Perpetrators choose these drugs because they act rapidly, produce disinhibition and relaxation of voluntary muscles, and cause the victim to have lasting anterograde amnesia for events that occur under the influence of the drug" (R. H. Schwartz, Milteer & Le Beau, 2000). By contrast, the high therapeutic dosages used to treat anxiety or panic disorders consistently depress sexual activity. In a study of patients with bipolar disorder (previously known as manic depression) who were being treated with both lithium and benzodiazepines, half reported difficulties in sexual functioning, including anorgasmia (Ghadirian, Annable & Belanger, 1992), a complaint not found with lithium treatment alone.

A similar picture emerges from case reports related to the use of another class of tranquilizers, the barbiturates. These drugs were used as sedatives and anxiolytic agents before benzodiazepines were introduced into medical practice. Their main property is sedation and, in adequate dosages, anesthesia. Not surprisingly, barbiturates induce sexual disorders and anorgasmia. The GABA-A receptor also contains a ligand site that is specific for barbiturates, which exert their sedating effect (at least partially) by enhancing GABAergic activity. An increase in GABAergic activity by positive modulators of the GABA-A receptor is apparently accompanied by inhibition of sexual activity in both men and women. Thus, administration of gabapentin was reported to produce anorgasmia in a male subject (Brannon & Rolland, 2000).

Opiates

Research on opiates was initially related to the attempt to develop potent analgesics (pain suppressants) without addictive properties. More recently the main focus of research has been on opiates' abuse properties. In general, they tend to inhibit sexual response. However, when the opiate heroin is rapidly administered intravenously, which produces high drug levels in the brain, a transient, intensely pleasurable sensation is produced that some individuals describe as orgasmic (De Leon & Wexler, 1973; Mirin et al., 1980; Seecof & Tennant, 1986). This orgasm-like experience of the heroin "rush" may be produced by stimulation of the ventral tegmental area of the brain, a region containing dopaminergic neurons that project to the "reward centers" of the limbic area, including the nucleus accumbens. PET (positron emission tomography) studies found that the ventral tegmental area was activated in men during ejaculation (Holstege et al., 2003), and fMRI (functional magnetic resonance imaging) studies showed the nucleus accumbens area to be activated in women during orgasm (Komisaruk, Whipple, Crawford, et al., 2002, 2004). Since this dopaminergic brain system is activated both by opiate drugs and by orgasm in men and women, this parallel could help explain the intensely reinforcing nature of both orgasm and heroin in men and women. It is likely, however, that the addictive craving for opiates is due to a mechanism other than the acute orgasm-like effect.

In William Burroughs's classic novel *Naked Lunch* (1959), Dr. Benway comments on this topic: "If all pleasure is relief from tension, junk [heroin] affords relief from the whole life process, in disconnecting the hypothalamus, which is the center of psychic energy and libido. Some of my learned colleagues (nameless assholes) have suggested that junk derives its euphoric effect from direct stimulation of the orgasm center. It seems more probable that junk dampens the whole cycle of tension, discharge and rest. Orgasm has no function in the junky." Indeed, chronic use of opiates (e.g., morphine, heroin, methadone) impairs all phases of sexual response. These effects are largely due to the inhibitory effect of opiates on the brain and its control over the pituitary-gonadal axis (Mirin et al., 1980; Pfaus & Gorzalka, 1987). Thus, heroin acutely suppresses secretion of luteinizing hormone from the pituitary gland, which decreases testosterone secretion by the testes, which in turn may eventually decrease sexual activity.

A similar inhibitory effect on sexual response, including orgasm frequency, also occurs in women and men receiving methadone as a treatment for opiate dependence (Crowley & Simpson, 1978). Conversely, during withdrawal from opiates, there are reports of a rapid and intense increase in libido "with a vengeance" (Erowid, 2005).

Recently, buprenorphine has been introduced for the pharmacotherapy of opiate dependence. This drug, in contrast to methadone and other opiates, does not decrease testosterone levels and is less frequently associated with negative sexual side effects (Bliesener et al., 2005).

15

Herbal Therapies

Given the long history of the search for substances capable of stimulating libido, enhancing orgasm, or increasing sexual satisfaction, there is now a long list of herbs that are alleged to possess such properties. In Rowland and Tai's review (2003) of plant-derived and herbal approaches to the treatment of "sexual disorders," they found very few products that have been tested in humans in double-blind, placebo-controlled studies. However, a few herbs considered in folklore to be aphrodisiacs have been found, in objective studies, to have positive effects on some phases of sexual response.

Herbal products do not require a prescription, are not regulated by food and drug agencies in many countries, are easy to obtain, and in some cases are relatively inexpensive compared with prescription medications. We review only those herbal products that have been tested in humans in double-blind, placebo-controlled studies.

Yohimbine

Yohimbine, derived from the bark of the African yohimbe tree, has had a reputation since antiquity as a remedy for erectile problems. It was one of the first substances to be tested for the treatment of erectile dysfunction (ED). Yohimbine is an alpha-2-adrenergic antagonist that affects both the central and peripheral nervous systems. Its pro-erectile effects are believed to be due, in part, to the inhibition of adrenergic receptors innervating genital tissue (Rowland & Tai, 2003). In a double-blind, placebo-controlled study, about 20 percent of men experienced strong positive effects on erection and 40 percent reported partial improvements with 15 to 30 mg/day of yohimbine (Morales et al., 1987; Rowland, Kallan & Slob, 1997). For women, the results are not as clear, with possible positive effects counteracting antidepressant-induced anorgasmia (Segraves, 1995). Yohimbine also disinhibits norepinephrine release, resulting in an "aphrodisiac" effect (E. Hollander & McCarley, 1992). The side effects of yohimbine include headache, sweating, agitation, hypertension, and sleeplessness. Its use is contraindicated for women and men who have certain cardiovascular, neurological, and psychological problems (Rowland & Tai, 2003).

Ginkgo Biloba

There is some evidence that extracts from the leaves of the *Ginkgo biloba* tree (also known as the maidenhair tree) have a significant effect on stimulating sexual response. Extracts of its dried leaves are available over-the-counter in many countries. Ginkgo biloba exerts vasodilating effects through two mechanisms: generation of nitric oxide and inhibition of platelet-activating factor. Several anecdotal and case reports suggest that ginkgo can facilitate sexual arousal in men and women. Ginkgo extracts have also been tested for their capacity to counteract the impairment of sexual response produced by a variety of antidepressants, including selective serotonin reuptake inhibitors (SSRIs), monoamine oxidase (MAO) inhibitors, and tricyclic antidepressants (TCAs) (Cohen & Bartlik, 1998). In that study, ginkgo was found to improve sexual function in 84 percent of the cases, more women responding favorably (91%) than men (76%). The beneficial effect of ginkgo was noted on libido, arousal, and orgasm.

Ginkgo extracts contain many potentially active agents, including flavonoids, glycosides, biflavones, and terpenlactones, that probably affect multiple neurotransmitters in the nervous system and could also be the basis for ginkgo's reputed beneficial effects on cognitive functions and memory. Its action on sexual arousal in men is most likely exerted through relaxation of the vascular smooth muscle of the corpora cavernosa, which increases blood flow into the penis. Therefore, to some extent, the ultimate effect of ginkgo, although not its mechanism of action, is comparable to that of sildenafil (Viagra). It is not clear how ginkgo affects women and their orgasms. While more comprehensive and well-controlled studies are required, it seems that ginkgo may be one of the few herbs that have aphrodisiac properties.

It must be noted that because of its "blood-thinning" effect, ginkgo is contraindicated for people taking anticoagulants such as heparin or warfarin (Coumadin), or anticoagulant medications such as aspirin or other nonsteroidal anti-inflammatory drugs. It also can interact adversely with some MAO inhibitors. People taking these medications should check the contents of mixed nutritional supplements for the presence of ginkgo.

Ginseng

Another ancient reputed aphrodisiac that has been validated by recent research is ginseng (Spinella, 2001). The term *ginseng* covers several related species of plants belonging to the genus *Panax*.

In addition to its reputation as an aphrodisiac, ginseng is also considered to be an "adaptogen," an agent that increases psychological and biological adaptation and resistance to stress (Sirpurapu et al., 2005). Ginseng contains steroids, saponins, and "ginsenosides." These components can affect acetylcholine, GABA, the monoamines, nitric oxide, and opioids. The clearest effect on sexual response is on sexual arousal, including penile erection via the nitric oxide pathway. Ginseng is more effective than trazodone for the treatment of certain sexual disorders, such as ED; the overall therapeutic efficacy for men with ED was found to be 60 percent for ginseng and 30 percent for trazodone (Choi & Seong, 1995).

ArginMax

ArginMax is sold as a nutritional supplement. The supplement for men contains Korean ginseng, American ginseng, ginkgo biloba, L-arginine (an amino acid), and various vitamins and minerals. The first, pilot study on the effects of ArginMax in men was an open-label study (i.e., both investigators and subjects receiving ArginMax knew they were receiving this supplement rather than a placebo) of 21 men with mild to moderate ED (Ito et al., 1998). The men were evaluated over a four-week period while taking the supplement. The results, based on a sexual function questionnaire, demonstrated an 89 percent increase in the ability to maintain an erection during intercourse, a 75 percent increase in satisfaction with overall sex life, and a 20 percent increase in number of orgasms. The only side effect was headache in 5 percent.

A subsequent randomized, double-blind, placebo-controlled study (i.e. neither the persons administering nor the subjects randomly selected to take ArginMax or placebo knew which they were receiving) included 52 participants with mild to severe ED (Ito & Kawahara, 2006). After four weeks, 84 percent of the subjects receiving ArginMax improved in their ability to maintain an erection during intercourse, compared with 24 percent in the placebo group; 80 percent reported improved satisfaction with overall sex life, compared with 20 percent for the placebo; and 36 percent reported an increase in orgasmic function, compared with 24 percent for placebo.

ArginMax for women is sold as a nutritional supplement that contains L-arginine, Korean ginseng, ginkgo biloba, various vitamins and minerals, and damiana (a progesterone-like substance derived from the plant *Turnera diffusa*). A double-blind, placebo-controlled study was conducted with 77 women; 34 received ArginMax and 43 received a placebo. After four weeks, 74 percent of the women who were taking ArginMax reported improved satisfaction with their overall sex life, compared with 37 percent in the placebo group. Significant improvements in the ArginMax treatment group were also reported in sexual desire, reduction of vaginal dryness, frequency of sexual intercourse, and orgasm. No side effects were noted in either group (Ito, Trant & Polan, 2001).

Zestra for Women

Many products are available over-the-counter for use as a lubricant to enhance sexual pleasure for women. Only one of these products, Zestra for Women, has been studied and the findings reported in a peer-reviewed journal. Zestra for Women is a natural botanical oil sold for the purpose of enhancing female sexual pleasure and arousal. It is applied directly to the vulva. Zestra contains borage seed oil, evening primrose oil, *Angelica* root extract, *Coleus forskohlii* extract, ascorbyl palmitate, and *dl*-alpha-tocopherol. A randomized, double-blind, placebo-controlled, cross-over study was conducted to evaluate the efficacy and safety of this product (D. M. Ferguson et al., 2003). (A cross-over experimental design involves administering the presumed active substance to one group and placebo to the other group, and then, after a period of time sufficient to assess the effects, reversing the treatments of the two groups and again affording sufficient time to assess the effects.

Consequently, over the entire course of the study, all participants receive both the presumed active substance and the placebo, although in a different sequence.) Ten women diagnosed with female sexual arousal disorder and ten women who did not have this diagnosis comprised the study population. The sexual arousal disorder group showed a significantly increased frequency of orgasm. All other variables tested—levels of desire and sexual arousal, satisfaction with level of sexual arousal, genital sensation, sexual pleasure, and enhancement of sexual experiences—showed some improvement with Zestra. Three of the women reported a single incident of mild genital burning sensations lasting five to thirty minutes after use of Zestra.

In addition to the herbal products described above, *Muira puama* extract, saw palmetto, *Tribulus terrestris* extract, and maca have also been associated with sexual function, but these have not been studied systematically in adequately sized populations of men and women (Rowland & Tai, 2003).

As a general precaution, health care providers and the general public should not recommend or use sexual health products unless they have been shown to be safe and effective in adequately controlled human clinical trials. Women are advised not to use products tested only in men, and vice versa.

16

Hormones and Orgasm

All reproductive processes, including sexual behavior and orgasm, are strongly influenced by the action of hormones on the central nervous system. Reciprocally, the central nervous system controls the endocrine glands that produce those hormones (Ferin, 1983). Several hormones of different chemical classes (proteins, peptides, and steroids) regulate the various phases of sexual behavior in mammals, including humans.

Sexual behavior depends on the coordinated functioning of a chain of hormones produced by the brain-pituitary-gonadal axis (McCann & Ojeda, 1996). In describing this brain-pituitary-gonadal axis, we select as an entry point the small neurons in the brain, mainly in the hypothalamus (Schally, 1978). The axon terminals of these neurons do not end in typical synapses.

Typical synapses are the functional microscopic gaps at which a neuron releases its neurotransmitter, which then stimulates or inhibits an adjacent neuron. By contrast, in the brain-pituitary-gonadal axis, the neuronal axon terminals end on a specialized net of blood capillaries, termed a "portal system," at the site where the pituitary gland is attached to the base of the brain. This is an unusual type of blood vessel because it has a capillary bed at both ends—one end in the brain, in connection with the neuron terminals, and the other end a very short distance away, in the anterior pituitary gland, in direct proximity with the cells that secrete pituitary hormones. Almost all other blood vessels—arteries and veins—have a capillary bed at only one end of the vessel, where it delivers blood to or removes blood from the vicinity of individual body cells. The other end of the vessel connects ultimately to the output or input of the heart, in the form of an open-pipe-like structure, not a capillary bed.

The unique function of the portal system between the brain and the anterior pituitary is to accept the neurochemicals released by the neuron terminals into the capillary bed in the hypothalamus and trans*port* (hence *portal*) them the very short distance (millimeters), directly and precisely, to individual pituitary gland cells via the capillary bed in the gland. Thus, this hypothalamo-pituitary portal system is an ultra-efficient structure for directly and precisely conveying extraordinarily small quantities of highly specific neurochemicals to the relatively few pituitary gland cells that use them, with the neurochemicals in highly concentrated—because highly localized—amounts, avoiding dilution to ineffectual levels by the whole-body bloodstream.

Certain of the neurochemicals released into this portal system by the specialized hypothalamic neurons are decapeptides, molecules with a structure like that of a charm bracelet, consisting of ten amino acids (each a "charm"). Gonadotropin-releasing hormone (GnRH) is a decapeptide synthesized in hypothalamic neurons and transported to the anterior pituitary gland. There, the GnRH stimulates specific pituitary cells to secrete two hormones known as "gonadotropins": follicle stimulating hormone (FSH) and luteinizing hormone (LH). Their function is to stimulate the gonads (testes and ovaries) to develop the gametes (spermatozoa and ova) and produce the sex steroid hormones. LH controls the synthesis and release into the blood of the hormones produced by the ovaries and testes, including estrogens, androgens, and progestins (there are several chemically related hormones of each type—e.g., estradiol, estriol, and estrone are estrogens—which have slightly different chemical structures and different potencies). These steroids are produced not only by the gonads but also by the adrenal cortex and even by brain tissue (in which case they are termed "neurosteroids"). Although it is commonly thought that women preferentially produce estrogens and men androgens, both types of sex steroids are produced by both sexes.

Hormones act through different cellular mechanisms, all of them involving a specific receptor (Mendelson, 1996). Some hormones (e.g., GnRH) act through receptors located on the cell membrane. The receptors typically deliver the hormone message to the interior of the cell via second messengers—for example, cyclic adenosine monophosphate (cyclic AMP, further abbreviated as cAMP). The action exerted by these hormones is usually of short latency (i.e., takes place after only a short delay) and relatively short duration. By contrast, steroid hormones (e.g., testosterone) penetrate freely into the cell and there bind to an intracellular receptor, which first combines with the hormone and then interacts with the genetic elements (the genome) of the cell to activate specific regions of DNA in the process termed "transcription." Transcription produces messenger RNA (mRNA), which then proceeds to synthesize specific proteins—the enzymes, receptors, transporters, and structural proteins that are the active elements and building blocks of the various cellular structures.

This so-called genomic action of steroid hormones takes many minutes to develop just the first step—protein synthesis. The overt response to steroid hormones (e.g., the growth of an organ) can take several days before it can be detected. Thus, testosterone administration restores sexual behavior or induces prostate growth in castrated male animals, but with a normal latency of several days.

Orgasm in humans, either the peripheral (i.e., involuntary visceral and voluntary muscular) component or the conscious experience, is a relatively brief event, averaging at most about 25 seconds in men and 17 seconds in women (Bohlen, Held & Sanderson, 1980; Bohlen et al., 1982; Carmichael, Warburton, et al., 1994; Mah & Binik, 2001). Both components of orgasm result from a brief activation of neural circuits, resulting from the excitation and inhibition of different groups of neurons through a variety of neurotransmitters, including dopamine, serotonin, and noradrenaline. It is clear that gonadal steroids, which act with long latencies, do not directly participate in the production of orgasm. Yet, an insufficient production of sex steroids (as in some pathological conditions or following surgical removal of the testes or ovaries) often results in anorgasmia and a decrease in libido in both sexes.

Sex Steroids in Males

Evidence of the importance of the gonads for sexual behavior is one of the most ancient biological observations. Castration of domestic animals was performed in ancient times and the behavioral effects of this operation were noted. Reference to human castration exists in Egyptian texts, and the use of this operation as punishment was established in Assyrian law as early as 1500 BC. A clear relationship between sexual desire and the presence of the testes was well-recognized in antiquity, as evidenced by the teachings of Valerius, an African Christian philosopher who advocated that the only safe way to achieve chastity was by castration—a common practice among his followers. Also known since ancient times was that some eunuchs retained their ability to engage in sexual intercourse. Indeed, in countries east of the Mediterranean, eunuchs were allowed to marry. And some Roman ladies, as mentioned by the satiric poet Juvenal, looked for the sexual favors of the infertile eunuchs.

Thus, it was recognized that although castration in men tended to decrease sexual behavior, there was great individual variability in the effect of the procedure. Factors such as age and previous sexual experience were recognized as playing a role in the survival time (i.e., persistence after castration) of libido and the capacity to experience orgasm.

The effect of castration on the sexual activity of both men and animals reveals that sexual desire and ejaculation-orgasm are different processes that are most likely integrated at different brain sites. Ejaculation-orgasm is most critically dependent on testicular secretions and usually disappears earliest following castration in animals. One of the first systematic studies on the dissociation of the various components of sexual behavior was conducted by Stone (1932) in the rabbit. He noted that ejaculation (or the behavioral pattern associated with ejaculation) disappeared rapidly (in 8 to 53 days) following castration, whereas mounting the female and performing pelvic thrusting persisted for several months more. In rats, it is also well established that the capacity for ejaculation is the first component of sexual behavior to disappear after castration (Stone, 1938). A large number of reports on primates indicate that in adult, sexually experienced male rhesus monkeys, castration produces a gradual decline in sexual behavior. Ejaculation is the first behavior pattern to decline—after two to four weeks following castration—whereas mounting behavior persists for longer periods (Michael & Wilson, 1973).

Although the relationship between the testis and sexual behavior was noted thousands of years ago, only much later was the behavioral change recognized as being linked to the absence of a substance produced by the testis. Théophile de Bordeau proposed the concept of "hormones" in the eighteenth century, but the first hormone—named "secretin"—was not discovered until the beginning of the twentieth century, by Bayliss and Starling (1902).

The idea that a substance secreted by the testis is essential to male sexual behavior, as well as to maintaining the physical characteristics of the male, originated from the studies of the German proto-endocrinologist A. A. Berthold in the mid–nineteenth century. This pioneer of endocrinology found that castration of roosters resulted in the atrophy (withering) of those tissues characteristic of the male (e.g., the prominent red crest and wattle of the head and throat) and in the loss of their behavioral repertoire, such as aggressiveness and sexual drive toward hens. Transplants of testicular tissue to castrated roosters not only corrected the tissue alterations induced by castration but restored the typical sex behavior. In 1910, E. Steinach extended this observation to mammals when he reported that if male rats castrated during infancy were implanted with testicular grafts, they showed normal copulatory behavior in adulthood (Steinach, 1940).

An intrepid pioneer of this type of studies in men was the nineteenth-century French physiologist Brown-Séquard. Based on Berthold's observation in roosters, he transplanted testicular tissue from a monkey to his own arm and claimed a positive effect on his well-being and his sexuality. The idea that testicular grafts could rejuvenate old or weak men, even to the point of restoring their sexuality, was championed by the Russian physician Serge Voronoff. He used testicular grafts from monkeys in his notorious attempts to fight age and sexual decline in some of his male patients. Although it has been said sarcastically that the only impact of the Voronoff procedure was to ensure the chastity of the monkeys, some recipients of the testicular grafts claimed positive effects on their sexual life. The possibility that testicular grafts could secrete sufficient amounts of hormone to restore sexual behavior was initially shown by Thorek (1924), in sexually inactive primates. He reported that testicular grafts from monkeys restored sexual activity in castrated individuals of the same species.

The availability of testosterone (the main androgen secreted by the testis) some years later led to studies confirming that, as proposed by Steinach (1940), androgens "eroticize" the male brain and stimulate all components of sexual behavior in a large diversity of species, including humans. In men, many studies in the first half of the twentieth century found that treatment with testosterone or some of its synthetic derivatives (e.g., testosterone propionate, methyltestosterone) increased or restored sexual response in men who were castrated or hypogonadal (Beach, 1948). In those studies, the occurrence or absence of ejaculation-orgasm was rarely mentioned.

There is no doubt that testosterone facilitates sexual behavior in men and that it is the main direct factor involved in the initiation of sexual interest at puberty. However, uncertainties and problems remain regarding just how important a role testosterone plays in human sexuality. For example, there are the intriguing observations that, in men: (1) testosterone is not *essential* to sexual response; (2) there are sources of androgen other than the testes; (3) androgens can increase sexual behavior above prior levels in eugonadal men (men with intact, normally functioning gonads); and (4) testosterone can be converted to estrogens (by a process termed "aromatization," a type of "biotransformation"), which can, surprisingly, stimulate *masculine* sexual behavior.

Extratesticular Androgen and Sexual Behavior

There is some disagreement about the indispensability of testicular androgen for the expression of sexual behavior in men. This is based on the observation that some men who have been castrated still experience sexual response, even culminating in orgasm, many years after the surgery. Substantial individual variability in the effect of castration on sexual behavior is also observed in nonhuman primates. Phoenix (1973) found that five out of ten castrated male rhesus monkeys still showed penile intromissions one year after the operation. Loy (1971) reported that a male rhesus castrated seven years earlier was still capable of displaying mounts, intromissions, and the ejaculatory pause (the temporary sexually inactive "refractory" period immediately after ejaculation). One of the obvious explanations for these large differences in the retention of sexual behavior by castrates is significant individual genetic differences in sensitivity to androgens or other sexual behavior–stimulating hormones in men and other primates. Although eugonadal adult men show a wide variability in testosterone plasma levels, these variations are uncorrelated with individual differences in sexual activity (Schiavi & White, 1976).

Studies in animals have revealed significant individual variations, probably due to genetic factors, in the amount of androgen required to maintain or restore copulatory behavior after castration. Therefore, it is possible that the sexual response that persists in men after castration is due to low levels of androgens that presumably can be produced by sources other than the testes (i.e., extratesticular sources). In addition to the testes, several tissues, both endocrine and para-endocrine (e.g., skin, liver; *para* denotes "atypical") can produce androgens. For example, the adrenal glands secrete substantial amounts of androstenedione and, particularly, dehydroepiandrosterone (DHEA) and its sulfate, all of which are weaker androgens than testosterone but can restore sexual behavior in castrated males of several species. Thus, in castrated rats, DHEA can induce all the behavioral components associated with copulation, including ejaculatory behavior (Beyer, Larsson, et al., 1973).

It is well established that brain tissue itself can produce "weak" androgens, including DHEA and its sulfate; thus, these hormones qualify as "neurosteroids" (Hu et al., 1987). Production of DHEA by the brain is completely independent of testicular or adrenal gland androgen secretion. Brain cells can generate high concentrations of this steroid in certain androgen-responsive brain areas. The role of neurosteroids in sexual behavior is as yet unknown.

The notion that sexual response in men is independent of androgens has been suggested by some authors who claim that in the course of evolution, primates and men gradually became independent of hormonal factors while social and learning factors acquired more importance as modulators of sexual life. At present, questions remain open as to whether (1) extratesticular sources of androgen can play a significant role in maintaining sexual activity in men after castration, and (2) sexual activity can persist in the true total absence of androgens.

Testosterone in Hypogonadal Males

The importance of testosterone for the normal stimulation of all phases of male sexual response is indicated by the many studies in which this androgen has restored or improved sexual activity in individuals with weak or no sexual activity due to low plasma levels of testosterone. A good review of the initial studies, up to the 1940s, which used testosterone or methyltestosterone treatment to improve sexual response in castrated men, is provided in the excellent, still valid and valuable book *Hormones and Behavior* by Frank Beach (1948). Subsequent studies have extended and confirmed these initial observations. Recent research on androgen treatment of sexual disorders in men has concentrated on developing methods to deliver testosterone, comfortably and safely, in an amount comparable to that produced by the testes and adequate to restore normal plasma levels of testosterone.

"Old Minds Rejuvenated by Sex Hormones"

A socially significant question is whether androgen treatment is an effective therapy to ameliorate the symptoms of andropause (or "male menopause"—an increasingly obsolete term), which is observed in some older men and is characterized by depressed mood, negative thinking, muscular weakness, and sexual disorders. As mentioned above, this problem of age-related hypogonadism was investigated by Brown-Séquard in the nineteenth century, using his classical testicular transplant experiment, as well as by the attempts of Voronoff to rejuvenate elderly men with grafts of monkey testicles. Steinach and Peczenik (1936) were the first to successfully use testosterone treatment to restore sexual functions in older men. At about the same time, N. E. Miller (1938) used the catchy phrase "old minds rejuvenated by sex hormones" to describe experiments in which administration of testosterone propionate improved the sexual activity of older men, whereas placebo injections failed to do so.

Currently, testosterone therapy is the standard treatment for hypogonadal men who complain of sexual disorders, including anorgasmia. Recent research on testosterone replacement therapy has concentrated on developing methods or formulations to deliver the androgen comfortably and safely in amounts adequate to restore normal plasma levels. Intramuscular injections of testosterone esters with prolonged ("depot") action produced large variations in testosterone concentration in plasma, and this treatment has been replaced by transdermal testosterone patches or gels, which provide a relatively constant release of testosterone into the circulation. Two such formulations—AndroGel (Wang et al., 2004) and AA2500 (Steidle et al., 2003)—studied in a large number of hypogonadal patients including elderly men, improved most components of sexual response, including sexual motivation, erectile capacity, and enjoyment with partner. From these studies it is clear that testosterone corrects some of the symptoms of late-onset hypogonadism, which include reduction in both interest and enjoyment of sex.

However, a controversy arose recently between an expert panel of the Endocrine Society and an expert panel of the National Institutes of Health (NIH). The Endocrine Society panel recommended treatment with the testosterone gel (AndroGel) for men older than 50 years who complain of symptoms of testosterone deficiency.

The NIH panel recommended caution in the use of the treatment until more information on the side effects of testosterone treatment became available. Moreover, the NIH panel emphasized difficulties in establishing the diagnosis of testosterone deficiency based on measurement of blood levels of testosterone. This issue has complex implications, including financial, and consequently the controversy is not likely to be resolved soon (Kassirer, 2004).

No reasonable doubt exists about the importance of testosterone to maintaining sexual response or restoring it in hypogonadal men. Yet, the capacity of testosterone to be an "aphrodisiac" that could stimulate supranormal sexual response in healthy, eugonadal men is unresolved. Some studies have reported that testosterone treatment increases sexual activity and enjoyment in eugonadal men (Alexander et al., 1997). This conclusion is supported to a certain extent by studies showing a good correlation between plasma levels of testosterone or 5-alpha-dihydrotestosterone (DHT) and the intensity and enjoyment of sexual activity (Fox, Ismail, et al., 1972; Kraemer et al., 1976; Mantzaros, Georgiadis & Trichopoulas, 1995). In a study performed on singers, Nieschlag (1979) found that basses have higher testosterone levels and greater frequency of ejaculations than tenors. The problematic nature of these correlational studies is evidenced by the findings of another study (Kraemer et al., 1976), not on singers, that there was no significant correlation between individual testosterone concentration and orgasmic frequency.

Other studies claim that above a certain androgen level, attained in most cases by normal testicular secretion, testosterone administration does not further increase sexual activity or quality of orgasm. For example, in a recent double-blind, placebo-controlled study on the effects of androgen on the sexual response of young eugonadal men, elevating testosterone and estradiol levels for about two weeks by administration of a large single dose of testosterone undecanoate (a long-acting ester of testosterone) failed to produce a clear effect on sexual activity or mood (O'Connor, Archer &

Woo, 2004). These observations in humans agree with some, but not all, findings in rodents and other species showing that high, supraphysiological doses of testosterone usually do not produce "hypersexuality." Rather, it seems that each individual has a characteristic level of sexual activity, most likely resulting from genetic, cultural, and experiential factors, that cannot be superseded by administering excessive amounts of testosterone. Some early results by Beach (1942), however, indicated that very large doses of androgen in eugonadal rats induced a "hypernormal degree of sexual *aggressiveness* combined with a decrease in the specificity of the stimulus adequate to affect sexual arousal." "Hypersexual" behavior has been reported in mice and other species treated with high dosages of androgen (P. E. Smith & Engle, 1927). Therefore, it is possible that creating abnormally high levels of plasma testosterone by administration of the hormone may produce some changes in the components of sexual behavior in men.

Experience and Responsiveness to Androgen

Analyses of the behavioral effects of androgen treatment in species including rodents, rabbits, carnivores, cattle, and nonhuman primates reveal that many factors, in addition to genetic, influence the behavioral response to these steroids. The effect of prior sexual experience on sexual behavior is particularly well illustrated in a study by Rosenblatt and Aronson (1958). They found that persistence of sexual activity after castration in male cats was positively correlated with prior sexual experience. Social factors may also play an important role in the response to androgens in monkeys. Thus, in talapoin monkeys, castration abolished almost all sexual activity when the animals were tested twelve months after the surgery. Testosterone replacement therapy restored copulatory behavior in the top-ranking males, but subordinate males that received comparable testosterone treatment failed to copulate, although they were observed to masturbate to ejaculation (Dixson & Herbert, 1977).

Biotransformation of Testosterone

For many years, testosterone was considered to be the main testicular hormone responsible for maintaining the structure and function of the male sex organs and accessory organs, including the prostate and seminal vesicles. However, biochemical studies of the prostate and seminal vesicles revealed that testosterone alone did not maintain growth in these structures. Instead, the hormone's 5-alpha-reduced metabolite, 5-alpha-dihydrotestosterone, was responsible for the effect. This suggested that in some target organs, testosterone is just a "prohormone" (a metabolic precursor of the true active substance that undergoes a biotransformation) and DHT is the actual active hormone (Bruchovsky & Wilson, 1968). Testosterone is converted to DHT by the enzyme 5-alpha-reductase, which acts on a particular portion—known as "ring A"—of the testosterone molecule and of related androgens (delta-4,3-keto-androgens) such as androstenedione. There is a widespread distribution of 5-alpha-reductase in the brain, particularly in areas of the limbic system, including the hypothalamus, preoptic area, and amygdala.

The role of 5-alpha reduction—the conversion of testosterone to DHT—in the stimulation of male sexual behavior in mammals is still unresolved. In some species—for example, rat (Beyer, Vidal & Mijares, 1970) and rabbit (Beyer & Rivaud, 1973)—DHT has no stimulatory effect on sexual behavior, suggesting that 5-alpha reduction alone has no significant role in the behavioral action of testosterone in these animals. By contrast, the effect of DHT on the sexual behavior of monkeys and men is not clear. Phoenix (1974) administered DHT in large doses to castrated rhesus monkeys and found a significant improvement in sexual activity, including ejaculation. But when Michael, Zumpe, and Bonsall (1986) compared the effect of testosterone propionate and DHT propionate in doses that restored normal plasma levels in intact male rhesus monkeys, they found DHT ineffective in stimulating sexual behavior. In these monkeys, treatment with higher, supranormal doses of DHT still failed to restore ejaculatory behavior.

In men 55 to 70 years of age, administration of DHT in the form of a topical transdermal patch has been reported to improve several of the negative traits characterizing late-onset hypogonadism, such as sexual disorders and loss of energy and muscle mass (De Lignieres, 1993). This treatment did not affect prostate weight and had no detectable adverse effect on prostate function, as judged by serum prostate-specific antigen levels and prostate volume.

Thus, it seems that DHT may stimulate some aspects of sexual response in men. Further studies are required to determine whether these effects are produced by DHT alone or in combination with the estrogens that could also be synthesized by aromatization of androgens of adrenal or testicular origin in various tissues of the body.

Testosterone undergoes a second major type of biotransformation: it is aromatized—and thus converted to an estrogen—through the action of aromatase enzymes. These enzymes are present in many tissues, including the brain. The distribution of aromatases in the brain has been determined in several species including humans (Naftolin et al., 1975). Aromatase enzymes are located in brain areas that are involved in sexual behavioral and endocrine reproductive processes. Although only a small proportion of testosterone is converted to estradiol (less than 1%), this transformation is physiologically significant because of the high potency of estradiol in affecting behavior. Aromatization of testosterone seems to be of major importance in facilitating *male* sexual behavior, at least in some species. Thus, blockage of aromatization prevents the facilitatory effect of testosterone on the sexual behavior of castrated male rats (Moralí, Larsson & Beyer, 1977), providing strong support for this enzymatic process. Curiously, estradiol alone does not stimulate male sexual behavior in most species studied. In men, estradiol is known to *inhibit* libido and produce anorgasmia.

While neither testosterone per se nor estradiol alone are effective in stimulating sexual behavior in male rats, administration of otherwise ineffectively low doses of estradiol *in combination with* DHT to castrated male rats elicited intense sexual behavior, including the ejaculatory pattern (Larsson, Sodersten & Beyer, 1973). Results obtained in males of several species—dove, Japanese quail, sheep, deer, ferret, and rabbit—also support the idea that normal male sexual behavior depends on the interaction of an androgen (testosterone or DHT) with an estrogen.

In primates, the role of aromatization in sexual behavior is unresolved. Zumpe, Bonsall, and Michael (1993) found that fadrazole, an aromatase inhibitor that prevents conversion of androgen to estrogen, significantly decreased ejaculatory frequency in castrated, testosterone-treated cynomolgus monkeys. This suggests that testosterone aromatization is an important process for stimulating ejaculatory behavior in primates. The role of aromatization in men is still uncertain, because few systematic studies have been conducted to test this possibility.

Although estrogen antagonists such as clomiphene do not impair sexual activity in men, this antiestrogen was unable to prevent the stimulatory effect of estradiol and DHT in combination in castrated male rats (Beyer, Morali, et al., 1976). However, some clinical studies indicate that estrogen can act synergistically with androgen to stimulate sexual response in men. Carani et al. (2005) recently reported the case of a man with hypogonadism and low libido who failed to respond to treatment with either estradiol or testosterone, but when he received both hormones in combination, he experienced a great increase in libido and incidence of orgasm.

Antiandrogen Administration to Sex Offenders

There are cases in which a society considers it beneficial to decrease a man's sexual activity. Almost a century ago, surgical castration was used for this purpose, and while this procedure has now been replaced by chemical castration, interest in its use for sex offenders has been reactivated (e.g., Gandhi, Purandare & Lock, 1993).

Drugs that reduce testosterone levels or interfere with the effects of testosterone (i.e., antiandrogens) have been used for many years, with variable success, to treat men with sexual compulsive disorders, particularly some paraphilias (Money, 1970). Men with such disorders may exhibit sexual responses that are socially and legally unacceptable. Treatment with medroxyprogesterone acetate (Cordoba & Chapel, 1983), which decreases testicular testosterone production, or with antiandrogens such as cyproterone acetate that prevent the action of testosterone on its receptor, decreases sexual activity, both appropriate and inappropriate. This confirms the importance of normal plasma levels of testosterone for the activity of the neural circuits involved in all phases of sexual response in men.

More recently, antidepressants and antipsychotics known to induce anorgasmia in men and women have been introduced for the treatment of undesirable sexual activity in men. Some of these drugs induce the release of the anterior pituitary hormone prolactin. The mechanism of action of this inhibition is not clear—whether it is the effect of the prolactin or of the drug itself (independent of the prolactin) on the brain.

Sex Steroids in Females

In females of nearly all nonprimate species, ovariectomy (surgical removal of the ovaries) results in a decrease and eventually a complete loss of sexual behavior. In most primates, including anthropoid primates (e.g., rhesus monkey, chimpanzee, talapoin), ovariectomy (in women, referred to as oophorectomy) reduces sexual behavior, and estradiol treatment restores the behavior, thus supporting the importance of this estrogen in female sexual behavior. A remarkable exception is the marmoset, which continues displaying sexual behavior following ovariectomy (Dixson, 1998).

Although castration in men usually results in anorgasmia and a general decline in sexual response, early studies in women tended to minimize the effect of oophorectomy on sexual response (Daniels & Tauber, 1941; Filler & Drezner, 1944; Heller, Farney & Myers, 1944). However, perhaps because of more sophisticated psychological measurement methods, the questions asked, and several well-controlled studies using hormone therapy, more recent studies support a significant role of the ovaries in sexual response in women. Thus, Shifren et al. (2000) reported that the ovaries provide approximately half the circulating testosterone in premenopausal women. In a study of 75 women who were premenopausal when they underwent surgical oophorectomy and hysterectomy, these authors stated: "Despite estrogen therapy, many surgically postmenopausal women have decreased sexual desire (libido), activity, and pleasure and a decreased general sense of well-being. These symptoms are believed to result from the lack of ovarian androgen production . . . The scores for thoughts-desire, frequency of sexual activity, and pleasure-orgasm were lowest at baseline and increased in a [testosterone] dose-dependent fashion. With the testosterone dose of 300 micrograms per day, the increases in scores for frequency of sexual activity and pleasure-orgasm were significantly greater than those with placebo." Consistent with these findings, Braunstein et al. (2005), in a study of 447 healthy women, reported that "the 300 microgram per day testosterone patch increased sexual desire and frequency of satisfying sexual activity and was well tolerated in women who developed hypoactive sexual desire disorder after surgical menopause" (i.e., combined surgical oophorectomy and hysterectomy).

Nathorst-Boos, von Shoultz, and Carlstrom (1993), in a study of 101 women who had undergone oophorectomy and hysterectomy, also found that the women "complained about less pleasure from coitus, impaired libido and lubrication . . . [however] estrogens are of little value in treating these specific sexual dysfunctions." Furthermore, "roughly one-third of the women experienced impaired lubrication at intercourse irrespective of whether . . . [estrogen therapy] had been administered or not. . . [and] regarding libido . . . almost half of the . . . women experienced a deterioration." In further confirmation of the greater effectiveness of testosterone than estrogen in restoring sexual response after oophorectomy, another study found lower self-reported desire and sexual arousal in women who received either no hormone treatment or estrogen only, by contrast with three comparison groups of women who underwent no surgery, or hysterectomy without oophorectomy, or oophorectomy plus combined androgen-estrogen therapy (Bellerose & Binik, 1993).

It should be noted that sex steroids (estrogens, androgens, progestins) can be produced, directly or indirectly, by tissues other than the ovaries and testes. The participation of the adrenal cortex as a source of steroids capable of maintaining sexual response in women after bilateral oophorectomy has often been suggested. Indeed, estrogen present in the blood of postmenopausal women, and women who have undergone bilateral oophorectomy, results from aromatization of androstenedione to estrone. In various tissues, approximately 1 percent of the androstenedione is converted to estrone (McDonald, 1971; Schindler, 1975).

Loss of sexual desire and anorgasmia were reported in women who had undergone both bilateral oophorectomy and bilateral surgical adrenalectomy (Salmon & Geist, 1943). This suggests that the chemical signals stimulating sexual response originate at least in part from the adrenal glands (Waxenburg, Drellich & Sutherland, 1959; Waxenburg et al., 1960). An important caveat is that the poor health of the women in this study (most having cancer and/or polycystic ovaries) must be taken into consideration in interpreting the findings.

The role of adrenal androgens in primate female sexual behavior has been studied either by adrenalectomy or by suppressing adrenal secretion by the administration of dexamethasone. Most of these studies revealed a decline in sexual activity, which could be restored by administering testosterone or androstenedione (Everitt & Herbert, 1971; Everitt, Herbert & Hamer, 1972). After natural menopause, when androgens are mainly produced by the adrenal cortex, plasma levels of androgen are positively related with sexual interest (Leiblum et al., 1983; McCoy & Davidson, 1985).

Menstrual Cycle, Sexual Behavior, and Orgasm

The menstrual cycle is, from an evolutionary point of view, a recent addition to the reproductive biology of mammals. Menstruation appears consistently only in Old World monkeys, apes, and women. A few New World monkeys, such as *Cebus apella,* menstruate (Nagle & Denari, 1983), but the quantity of blood loss is much less than in Old World monkeys. Some New World monkeys, such as the squirrel monkey, have a clear estrous cycle in which copulation occurs on a single day, when the female accepts mounting by the male. On that day, the LH peak occurs and induces ovulation. Although not as closely restricted to the periovulatory period (the period around the time of ovulation) as in the squirrel monkey, other New World monkeys have somewhat longer periods of sexual activity during which the female shows proceptivity (behavioral patterns performed by the female, such as "invitations" to copulate) and sexual receptivity (for a review, see Dixson, 1998).

The most detailed studies on the relationship between ovarian hormone levels (estradiol, testosterone, and progesterone) and sexual behavior in primates have been made in the New World marmoset (Kendrick & Dixson, 1983). In this species, a clear correlation was found between testosterone levels and proceptivity during the periovulatory phase. Proceptivity practically disappeared during the later, luteal phase, the phase of the reproductive cycle, just after the ova are released from the ovarian follicles, when the corpora lutea form in the vacated follicles and begin secreting progesterone (which stimulates the development of the uterine wall in preparation for implantation of fertilized ova). During this phase, when plasma progesterone levels are high, the females tended to refuse mounting attempts by the male, but they nevertheless copulated throughout the cycle.

These results contrast with those obtained in the more "primitive" primates—the prosimians, such as lemurs—which restrict their sexual behavior to the periovulatory period. New World monkeys, although clearly stimulated to display proceptivity by testosterone or estradiol, may copulate at other stages of the reproductive cycle. These findings may be interpreted as an indication that sexual receptivity in New World monkeys is less dependent on ovarian hormones than in prosimians and, of course, in nonprimate mammals, which tend to have limited periods of sexual receptivity (or estrus) that is more rigidly controlled by hormones.

Among the Old World monkeys, catarrhine primates (e.g., *Macaca mulatta,* rhesus monkeys) copulate more frequently during the follicular phase of the menstrual cycle (Wallen et al., 1984), when the ova are growing, but substantial variability exists among studies performed in the field and in the laboratory. Indeed, macaques can continue copulating during pregnancy, as do the female anthropoid apes and women. Again, although females copulate throughout the menstrual cycle, peaks of sexual activity and particularly proceptivity occur at the late follicular phase (before ovulation) and in the periovulatory period (Michael & Welegalla, 1968; Goy & Resko, 1972; Wallen et al., 1984; for a review, see Dixson, 1998). A similar behavior exists in most of the great apes. Female chimpanzees mate at any time of the menstrual cycle, but proceptive behavior occurs mainly during the follicular and periovulatory phase. It seems that the female great apes will accept males, particularly vigorous and aggressive ones, during the follicular as well as the luteal phase, but proceptive behavior, which is considered to be analogous to sexual desire in women, is mainly restricted to the periovulatory period. An interesting exception is the gorilla, in which copulation is restricted to two to three days at midcycle. During this period, the female is highly proceptive and shows swelling of the labia of the external genitalia, indicative of the action of estrogen on those tissues (Nadler, 1975).

As is clear from analyses of sexual activity during the menstrual cycle, nonreproductive mating occurs not only in humans but in nonhuman primates. Hormones clearly affect some components of sexual behavior—such as female attractiveness to males, consisting of changes in the female external genitalia—that are restricted to the follicular phase. Similarly, proceptivity also tends to concentrate in the periovulatory period. Estradiol and testosterone levels, which peak at this time, may participate in stimulating proceptivity. Although not specifically studied, it is possible that during this periovulatory period, in which females actively seek copulation, orgasmic responses are more frequent, while during the luteal phase, in which females do not seek but will engage in copulation, orgasmic responses are less frequent.

Long before measurement of plasma levels of sex hormones became possible, it was evident that women engagc in sexual intercourse and experience orgasm throughout the menstrual cycle. Some researchers believe that the highest incidence of sexual activity occurs around the time of ovulation, when sex steroid secretion is highest, a situation similar to that in nonhuman primates. However, there is a lack of consensus among the various investigators who have studied this issue.

K. B. Davis (1929), in his pioneering study, concluded from an analysis of 2,200 women that the intensity of sexual desire occurs in two peaks: one before and another after menstruation. This conclusion was shared by Greenblatt (1943), who claimed that some women become "nymphomaniac" just before menstruation. An analysis of all studies available on the subject up to 1980 found that peaks of sexual activity in women have been reported primarily at midcycle (8 studies), premenstrually (17 studies), postmenstrually (18 studies), and during menstruation (4 studies) (Schreiner-Engel, 1980; Schreiner-Engel, Schiavi, et al., 1981). We can draw some conclusions from this seemingly inconsistent picture. The safest one is that women engage in sexual activity at any stage of the menstrual cycle and therefore under different hormonal conditions—that is, high or low levels of estradiol and presence or absence of progesterone.

That endogenous ovarian hormones do not seem to play an essential ("obligatory") role in women's sexual response is also supported by the observation that at menopause, when levels of estradiol, progesterone, and testosterone decline, there is no particularly dramatic effect on sexual activity. Some other nonhormonal factors such as health, sexual partner, and mental health were found to have a "greater impact on women's sexual functioning than menopause status" (Avis et al., 2000; Farrington, 2005). Moreover, during pregnancy, when plasma progesterone levels are high, women commonly engage in sexual behavior and experience orgasm.

A hormonal influence on sexual response in women is suggested by evidence that both sexual desire and sexual enjoyment, including orgasm, occur more frequently during the periovulatory phase than other phases of the menstrual cycle (Udry & Morris, 1968; Adams, Gold & Burt, 1978; Bancroft et al., 1983). These changes are probably related to the peak of plasma androgen levels before ovulation. In further support of this suggested role of androgen in women's sexual response, Persky, Lief, et al. (1978) found a relation between midcycle testosterone levels and frequency of orgasm, and Bancroft et al. (1983) found a similar trend for masturbation.

Estrogens

Estradiol is essential to the display of sexual behavior in female nonprimate mammals. In some species (cat, rabbit), estrogen alone facilitates sexual behavior, while in others (rats, mice), combination with progesterone is obligatory. Therefore, it is surprising that estradiol apparently does not play any significant role in women's sexual response, including orgasm.

Numerous studies have been conducted in which estradiol or other estrogens are administered to women, under various conditions: women with bilateral oophorectomy, eugonadal women who have low sex drive, or women who are anorgasmic. All these studies agree that there is no consistent effect of estrogen treatment on the various components of sexual response—sexual drive, sexual dreams, or occurrence and quality of orgasm (Sherwin, Gelfand & Brender, 1985; Utian, 1975).

Androgens

Women produce androgens in the ovaries and the adrenal glands, and these androgens may stimulate several components of sexual response. The ovary secretes testosterone in varying amounts depending on the stage of the menstrual cycle, with levels being highest at the time of ovulation. Several studies have shown that in premenopausal women there is a significant correlation between sexual activity, including frequency of orgasms, and plasma levels of testosterone. Moreover, a large number of studies of premenopausal and postmenopausal women indicate that testosterone administration improves all aspects of sexual response (Sherwin, Gelfand & Brender, 1985; Sherwin & Gelfand, 1987; Sherwin, 1988; Van Goozen et al., 1997). A serious drawback of testosterone therapy for sexual disorders in women is its virilizing effect, such as producing alopecia (baldness), hirsutism (increased body hair), or unwanted muscular development. Recent studies report that by carefully adjusting testosterone dosages, these negative side effects can be avoided. Thus, transdermal (skin patch) application of low doses of testosterone (150 to 300 micrograms) for women who had undergone bilateral oophorectomy and hysterectomy and reported impaired sexual function significantly improved sexual activity, including pleasurable orgasm (Shifren et al., 2000), without producing virilization or other undesired side effects. These women also received "conjugated" estrogen, which, by itself, had no effect on sexual response.

It is possible that androgens other than testosterone influence sexual response in women. The adrenal glands secrete high levels of dehydroepiandrosterone, particularly the sulfate form. A recent study found that women with low sexual desire, either pre- or postmenopausal, had consistently lower testosterone and DHEA levels than women with normal desire (Turna et al., 2005). S. R. Davis et al. (2005) reported that a low score for sexual responsiveness was associated with a low level of DHEA but not with the levels of testosterone or androstenedione. However, not all women with low DHEA had low levels of sexual activity.

These nontestosterone steroids are considered to be weak androgens, on the basis that they only weakly stimulate the prostate gland and seminal vesicles in men and do not induce virilization in women. However, findings in animals suggest that they may exert significant effects on brain functions related to cognition and emotionality. Moreover, DHEA can restore sexual behavior in ovariectomized rabbits and rats (Beyer, Vidal & Mijares, 1970). One study found a significant positive effect of DHEA administration on the sexual arousal of postmenopausal women (Hackbert, Heiman & Meston, 1998). It remains to be determined whether this is due to a direct effect of DHEA on brain structures related to sexual behavior or to an indirect effect resulting from the ability of DHEA to counteract the deteriorating effects of aging on nonbrain tissues. Administration of DHEA to premenopausal women had no clear effect on sexual arousal (Meston & Heiman, 2002).

Enzymatic Reduction of Testosterone

In women as in men, testosterone is rapidly metabolized to 5-alpha-DHT, a more active androgen than testosterone. It is not known whether such "ring A–reduced" metabolites of testosterone are involved in facilitating libido in women. A recent report supporting this possibility describes the case of a woman with life-long absence of sexual desire. The only abnormality found after a thorough endocrine study was a low level of 5-alpha-DHT. Administration of a 5-alpha-DHT gel to the woman's vulva stimulated her sexual desire and her capacity to be sexually aroused (Riley, 1999). A similar, much earlier study reported that the application of a different ring A–reduced androgen—androsterone—to the vagina increased sexual desire in sexually hypoactive women (A. S. Parkes, personal communication).

Several studies indicate that estrogen administration augments the positive effects of testosterone (A. Davis et al., 2003; Bachmann & Leiblum, 2004). Sarrel (1999) concluded that hormone therapy with estrogen plus androgen provides a greater improvement in both psychological (depression, fatigue) and sexual (decreased libido, anorgasmia) symptoms than does estrogen alone for surgically and naturally menopausal women. Moreover, in women with surgically induced menopause, the combined administration of androgen and estrogen effectively restored their previous levels of sexual desire, a response not shown by treatment with estrogen alone. These findings suggest that testosterone and estradiol can act synergistically to facilitate orgasm and other aspects of women's sexual response, as has also been found in some animal species.

Progesterone

Progesterone is the sex hormone secreted in largest amounts by the ovaries, particularly during the luteal phase of the menstrual cycle. It is also secreted in substantial amounts by the adrenal glands, in both women and men. Early studies analyzed various characteristics of women's mood and behavior in relation to plasma levels of progesterone. There is evidence that the negative mood changes observed in the premenstrual period in some women ("premenstrual syndrome") are related to a decline in plasma levels of progestins at that phase of the menstrual cycle.

In nearly all mammals, progesterone has an important role in the modulation of sexual behavior in females. In rodents it triggers estrous behavior in females previously treated with estrogen. Progesterone controls the timing and duration of sexual responsiveness; following its initial stimulatory effect, the hormone produces a decrease in estrous behavior. In other species, such as rabbits, that do not require progesterone to facilitate lordosis (elevation of the rump and vaginal opening for mating), progesterone exerts a clear inhibitory effect on sexual behavior.

Therefore, it is surprising that there is almost universal agreement that progesterone treatment does not have a marked effect on the sexual response of either premenopausal or postmenopausal women (Persky, O'Brien & Kahn, 1976; Schreiner-Engel, Schiavi, et al., 1981) or of primates in general (Dixson, 1998). Although some researchers have reported changes in libido in women taking oral contraceptives that contain synthetic progestins, the direction of the changes (positive or [mostly] negative) and the proportion of affected subjects vary considerably among the reports. Moreover, most contraceptive preparations contain both estrogen and progestins, making it difficult to ascribe reported changes to progestins per se. However, there is no solid basis on which to claim an effect of progesterone on orgasm. For example, Caruso et al. (2004) studied the effect of taking low-dose oral contraceptives containing ethinyl-estradiol and the progestin gestodene. They reported some effects on libido and sexual arousal, presumably due to low androgen levels, but the frequency of orgasm did not change during the nine months of oral contraceptive use.

17

Mechanism of Action of Sex Steroids

The sex steroids produce an initial cascade of cellular processes that is basically the same in the brain as in the sex steroid–sensitive peripheral organs, such as the uterus and prostate. Steroid hormones interact with receptors inside their target cells. They penetrate into the cell, freely crossing the cell membrane, and bind with a specific intracellular (often intranuclear) receptor protein. The highly specific interaction between hormone and receptor triggers a series of events culminating in gene (DNA) transcription and the synthesis of proteins. This process may take a few hours, but its effect on behavior persists for days—that is, long after the steroid has disappeared from both the cell and the bloodstream. Interference with this process by administration of an antihormone (e.g., the antiandrogen cyproterone acetate) blocks the behavioral effect.

Testosterone and the other steroid hormones do not trigger the pattern of neuronal activity leading to ejaculation or orgasm. Indeed, plasma levels of testosterone do not increase during arousal or orgasm in men (Krüger, Haake, Hartmann, et al., 2002; Krüger, Haake, Chereath, et al., 2003). Rather, testosterone and other androgens act by building up and maintaining the functionality of the diverse neural circuits involved in these sexual behavior patterns. This explains why the latencies to restore sexual response in hypogonadal or castrated subjects are quite extended (several days to weeks), most likely involving trophic ("nutritive") actions and the synthesis of structural proteins.

Steroid hormones modulate brain processes such as pituitary secretion and reproductive behavior by two main types of brain processes, which Frank Beach (1975) termed "organizational," differentiating the developing brain toward feminine or masculine behavioral and physiological patterns, and "activational," triggering the expression of feminine or masculine behavioral and physiological patterns in the adult. The organizational processes occur during a restricted phase of brain development, during the perinatal period, and result in the sexual differentiation of the brain, in which the brain develops neural components in females that are different from those in males.

Studies performed mainly in rodents have revealed that the development of a neural circuit for the production of male sexual response, including ejaculation, depends on the secretion of testosterone by the fetal testis during the prenatal organizational period. Testosterone and its metabolites (including estradiol) induce a variety of structural changes in several brain and spinal cord regions and determine the number of neurons and the patterns of connections of specific brain nuclei (clusters of neurons). This organizational effect becomes particularly evident when testosterone is administered to neonatal female rats within two days after birth. As a result of this brief early treatment with testosterone, the female rat, when an adult, not only fails to ovulate but displays toward other females the behavioral pattern associated with ejaculation in male rats (Moralí, Larsson & Beyer, 1977).

Some human brain structures are sexually dimorphic, differing in physical structure between males and females (Le Vay, 1993). Particularly relevant for sexual response may be differences in the size and number of neurons found in the interstitial nuclei of the anterior hypothalamus. Swaab and Fliers (1985) identified a sexually dimorphic nucleus (larger in men than in women). The difference develops between the age of 4 years and puberty and is due to a higher rate of apoptosis (normally occurring neuronal cell death) in women than in men. Other studies have found other nuclei in this area that seem to be larger in men than in women (Le Vay, 1991, 1993). There is no information, however, on whether hormones participate in this process or on the functional significance of this difference.

Although self-reported descriptions of orgasm by women and men are indistinguishable, the physiological events in the two sexes are certainly different. Recent studies with PET (positron emission tomography) (Holstege et al., 2003) and fMRI (functional magnetic resonance imaging) (Komisaruk, Whipple, Crawford, et al., 2002, 2004) show regional differences in the pattern of brain activation during orgasm in men and women, which may reflect physiological, more than cognitive, differences. Thus, it is likely that sex differences in the neural circuits involved in orgasm in humans are produced by organizational actions of androgens or other sex steroids during brain development.

Following its organizational effect, exerted during the perinatal period, androgen secretion by the testis remains very low until puberty, when testosterone or its metabolites initiate the multiple manifestations of sexual behavior in males. Increases in testosterone, which is measurable in saliva, are closely related to the initiation of sexual activity, including the experience of orgasm (Halpern, Udry & Suchindran, 1998). The levels of free plasma testosterone (and/or its metabolites) correlate well with the frequency of sexual thoughts in adolescent boys (Halpern et al., 1994). Thus, the initiation of male sexual response at puberty, as well as its maintenance throughout life, is due to the *activational* effect of testosterone on previously *organized* neural circuits. Testosterone by itself may activate sexual desire, but it does not directly participate in stimulating the neural circuits involved in sexual arousal or orgasm. The role of testosterone in these processes is to maintain the functionality and capacity of these circuits to respond to the complex sensory activity provided by genital stimulation.

The role of testosterone in maintaining sexual response in all its phases, as well as the gradual deterioration of sexual response following testosterone removal, provides information about the unique characteristics of the underlying neural circuits. The androgen-dependent circuits require the continuous presence of testosterone or its metabolites to remain functional, much as a highway requires maintenance and repair. We have little information about the nature of the cellular events regulated by testosterone in the brain, although most likely they are trophic activities, involving axonal growth or "sprouting" of axon terminals. It is evidently through these processes that the circuits involved in orgasm are maintained and adjusted for optimal functioning.

We also lack information about the human brain structures on which testosterone acts to stimulate sexual response, including orgasm. Since testosterone most likely acts through androgen receptors and, via aromatization, through estrogen receptors, brain structures that have testosterone receptors and, probably, estrogen receptors are logical candidates to be involved in sexual response, including orgasm.

A great advance in the methodology to localize steroid receptors came from the synthesis of radioactively labeled steroid hormones, which allowed investigators to detect the binding of the hormone to its responsive tissues in various sites in the body, including the brain and tissues related to the genital system. Also made possible by new technologies was the purification of a protein that binds the steroid hormones in the blood and, later, the generation of antibodies to the steroid receptors. Using monoclonal antibodies to the proteins comprising the steroid receptors, researchers have been able to map the distribution of receptors throughout the brain and spinal cord. These maps provide clues as to the neural sites of action of the steroid hormones.

The brain structures that have sex steroid receptors are components of the limbic system, a part of the brain that developed, evolutionarily, before the highly specialized neocortex of humans and is related to the control of autonomic and hormonal processes. It seems that androgen, by binding to these receptors, triggers a variety of changes including structural and biochemical changes, which, in turn, alter the responsiveness of specific neural circuits. With these neural circuits adequately "eroticized" by the steroid hormones, the presence of appropriate sensory stimuli—such as a mate—will trigger the appropriate sexual response. The hormones, in the absence of specific sensory stimuli, induce sexual motivation or desire, leading to a series of overt behavioral actions that result in sexual interaction.

There is remarkable consistency in the distribution of androgen and estrogen receptors in the brain across mammalian species. Most of the brain neurons that have androgen receptors are located in the medial periventricular region of the hypothalamus and in the medial preoptic area. In primates (such as *Macaca mulatta,* rhesus monkeys), some forebrain areas, such as the amygdala and the bed nucleus of the stria terminalis, have androgen receptors (Bonsall, Rees & Michael, 1989). That these regions are involved in libido or sexual motivation is supported by research in rodents in which implants of testosterone in the medial preoptic area in castrated animals restore sexual motivation (E. R. Smith, Damassa & Davidson, 1977).

Taken in a broad perspective, the contrast is puzzling between the significant role of sex steroids in sexual behavior in female nonhuman mammals and the evident failure of estrogen to stimulate sexual response and orgasm expression in women (Schreiner-Engel, Schiavi, et al., 1989). All anthropoid apes that have been studied show dependence on estrogen for sexual activity. It is difficult to differentiate peripheral (e.g., genital organ) from central (e.g., brain) effects in monkeys. Estrogen acting at the external genitals (vulva and vagina) stimulates females' attractiveness to males, and there is good evidence that estrogen also acts in the brain to stimulate proceptivity and receptivity. This contrasts with the failure of even high-dose estrogen treatment in women to stimulate libido, sexual arousal, or orgasm (Salmon & Geist, 1943; Furuhjelm, Karlgren & Carlstrom, 1984; Nathorst-Boos, von Schoultz & Carlstrom, 1993).

Yet, there is little doubt that testosterone and perhaps other androgens exert an aphrodisiac effect in women, both pre- and postmenopausal. Recent well-controlled studies reveal that when added to chronic estrogen treatment, chronic testosterone administration (via skin patch) that restored plasma testosterone to normal levels significantly improved all aspects of sexual activity in women, including orgasm (Shifren et al., 2000). The effect of dihydrotestosterone on women's sexual response is not clear, but when it is combined with estrogen, the combination significantly stimulates all aspects of sexual activity. Similarly, estrogen, ineffective by itself, seems to act synergistically with androgen treatment to facilitate sexual response in women.

These findings suggest that contrary to what has been generally believed, estrogen may indeed play an important role in women's sexual response—not alone, but when combined with androgen. From the available evidence, we are led to conclude that both men and women use a combination of androgen and estrogen, *naturally* resulting from testosterone metabolism, to facilitate sexual response, including orgasm.

18

Nonreproductive Hormones in Orgasm

Hormones other than the sex steroids (androgens and estrogens) have been reported to influence sexual response, but their natural role is still debated. Some of these hormones, such as oxytocin and prolactin, are released in both sexes during sexual arousal and orgasm, and prolactin may play a role in postorgasmic behavior—in the "resolution phase" described by Masters and Johnson (1966). Here we discuss the effects on orgasm of adrenal cortical hormones, oxytocin, and prolactin.

Adrenal Cortical Hormones

The adrenal cortical hormones (e.g., cortisol), also known as corticoids, are steroids secreted by the adrenal cortex. They have not been well studied in relation to their possible role in sexual response and orgasm, but to the extent that mood and sexual desire are related, corticoids most likely play an important, though indirect, role in sexual response. For example, the increased secretion of corticoids that is characteristic of Cushing disease (which results from a benign tumor of the anterior pituitary gland) is frequently associated with depression and loss of sexual interest (Starkman & Schteingart, 1981). The decrease in libido is probably due not to a direct inhibitory effect of corticoids on the neural networks involved in sexual motivation but rather to the depression that accompanies the brain activity that produces hypersecretion of cortisol. High levels of cortisol in men decrease testosterone secretion by decreasing testicular responsiveness to luteinizing hormone (Sapolsky, 1985), which can lead to a decreased sexual response. A decrease in libido is usually associated with high levels of corticotropin, also known as adrenocorticotropic hormone (ACTH), which controls secretion of corticoids from the adrenal cortex. Decrease in libido due to hyperactivity of the pituitary-adrenal axis occurs in both sexes. Sexual arousal and orgasm are *not* associated with significant changes in corticoid secretion, thus reinforcing the idea that corticoids do not directly stimulate any aspect of sexual behavior.

Oxytocin

Oxytocin plays a role in penile erection and is released during orgasm. Of interest here is whether oxytocin affects the *perception* of orgasm. Oxytocin is a nonapeptide (a peptide containing nine amino acids) synthesized in both large neurons (in supraoptic and paraventricular nuclei) and small neurons (in tuberal nuclei) of the hypothalamus. The axons of some of these neurons end in close association with capillary blood vessels in the posterior pituitary gland (also called the "neural lobe of the pituitary gland"), rather than synapsing with other neurons. This unique anatomical arrangement allows oxytocin to be released directly into the bloodstream, as a hormone, from neurons in the brain—a process termed "neurohormone secretion." Other oxytocinergic neurons release oxytocin into the capillary bed of the portal vessels, which carry it a very short distance (millimeters) to a second capillary bed in the anterior pituitary gland, where oxytocin modulates the release of ACTH. In yet another anatomical arrangement, the axons of some oxytocinergic neurons synapse with neurons in the lower brainstem and at various points along the spinal cord, to modulate the excitability of these neurons. Thus, oxytocin has the rare distinction of acting as both a conventional neurotransmitter and a conventional hormone.

Stimulation of the nipple area by suckling or of the genital tract during childbirth provokes the release of oxytocin into the bloodstream. The subsequent action of oxytocin on the mammary myoepithelium produces milk ejection, and its action on the uterine smooth muscle produces uterine contractions. Copulation elicits oxytocin release in many mammalian species, including humans (Blaicher et al., 1999; Krüger, Haake, Chereath, et al., 2003).

Because oxytocin stimulates the contraction of the smooth muscle of some sex organs involved in ejaculation, it seems probable that this neurohormone directly contributes to the peripheral orgasmic response (Filippi et al., 2003). Some researchers have suggested that oxytocin may also be involved in the sensory experience of orgasm. A positive correlation between the intensity of orgasm and the concentration of plasma oxytocin has been found in both sexes (Carmichael, Warburton, et al., 1994), although this could represent a joint effect of a more intense orgasm releasing more oxytocin, rather than oxytocin having an effect on the brain to intensify orgasm. Some investigators reported, however, that administration of naloxone to men, a procedure that reduces plasma levels of oxytocin at orgasm, decreased pleasure from orgasm (Murphy et al., 1990). Again, this could be due to the naloxone reducing both the intensity of the perceived orgasm (perhaps via an effect on endogenous opioids) and the amount of oxytocin released, rather than the oxytocin having a direct effect on the perceived intensity of orgasm.

Oxytocin does not readily pass from the blood into the brain (i.e., across the blood-brain barrier). Oxytocinergic neurons that send their axons to other neurons (rather than to the neural lobe of the pituitary gland) may stimulate neurons of the brain circuit involved in the sensory experience of orgasm.

Release of oxytocin during coitus in women has been related to the facilitation of sperm transport, but its possible role in men, besides some participation in the peripheral orgasmic response, is uncertain. From animal studies, it is known that oxytocin participates in affiliative processes, and thus the release of oxytocin during orgasm could result in strengthening of affiliative responses between partners (Carter et al., 1992; Insel et al., 1993).

Prolactin

Prolactin is a protein hormone that is synthesized by cells in the anterior pituitary gland, from which it is released into the bloodstream. There are data to suggest that prolactin can influence some of the components of sexual response, including orgasm. Numerous studies have reported that a pathologically high level of prolactin in plasma—a condition known as hyperprolactinemia—is often associated with a decrease in both libido and sexual satisfaction in men and women (M. B. Schwartz, Bauman & Masters, 1982; Bancroft, 1984; Buchman & Kellner, 1984). Because some neuroleptic drugs and antidepressants increase prolactin secretion, it has been proposed that the impairment of sexual functioning (including anorgasmia) induced by such drugs is mediated by this process—that is, as a direct inhibitory effect of prolactin, a direct inhibitory effect of the drugs, or both.

The sexual behavioral symptoms of hyperprolactinemia are reversed when patients are treated with bromocryptine (Buchman & Kellner, 1984), a dopamine agonist that inhibits prolactin secretion and consequently decreases its level in plasma. Although these observations link abnormally high levels of prolactin with a loss of sexual desire, it is not clear whether secretion of prolactin normally modulates sexual response. Some researchers have reported that breast feeding, a stimulus that elicits prolactin release, tends to decrease sexual desire in lactating women (Kayner & Sager, 1983). However, the relationship may be only indirect, since the depressed mood that occurs in some postpartum women could be more directly related to the decrease in sexual desire.

Recent studies have demonstrated that prolactin secretion consistently rises in both men and women during orgasm (e.g., Kruger, Hartmann & Schedlowski, 2005). This endocrine response seems to be specific to orgasm, since sexual arousal is accompanied by only minimal, if any, increases in prolactin levels. Consistent with these findings, there is a report of a man capable of experiencing three orgasms without intervening postejaculatory "refractory" periods (time unresponsive to sexual stimulation) who failed to show any rise in prolactin following orgasm (for a review, see Krüger, Haake, Hartmann, et al., 2002). From these observations, it has been proposed that normally, the prolactin released at orgasm almost immediately "feeds back" on the brain circuits related to sexual motivation and inhibits sexual drive.

Although prolactin does not penetrate the blood-brain barrier, it does gain access to the cerebrospinal fluid, which bathes the brain and spinal cord, from where it might modulate neuronal activity in regions such as the medial preoptic area, the septum, and the hypothalamus; these brain structures have been related to the expression of sexual response. Prolactin receptors have been detected in the brain, thus supporting a possible site of action for a prolactin effect.

19

Atypical Orgasms

There are reports of a variety of orgasms that occur under atypical conditions, apparently independent of genital stimulation (which generates "typical orgasms"). A consideration of the conditions under which these "atypical orgasms" occur—the factors that stimulate them, the individuals who experience them, and the qualities of the orgasms—is instructive in understanding the nature of orgasm. When we consider the cases described here, *nongenital orgasm* is not an oxymoron.

During Dreaming Sleep

Several lines of evidence demonstrate that the brain can generate an orgasm independent of genital sensory activity in both men and women. Physiological changes (vaginal blood flow, heart rate, and respiration rate) were measured in a sleeping woman while she had an orgasm in a dream, which she described when she awoke (Fisher et al., 1983). During the orgasm, her heart rate increased from 50 to 100 beats per minute and her respiration from 12 to 22 breaths per minute, and she had a "very marked" increase in vaginal blood flow. The vaginal blood flow showed cyclic episodes of vascular engorgement (the equivalent of penile erections in men) during REM (rapid eye movement) sleep periods. The vaginal vascular engorgement tended to occur with a frequency equal to that of penile erections in men (i.e., during 95% of REM periods). However, during non-REM sleep her vaginal responses occurred with greater frequency than penile erections in men; they were shorter in duration and less tightly linked to the REM periods.

In Fisher et al.'s case study, the brain activity that generated the woman's experience of orgasm was not a response to genital stimulation, but the brain activity generated output to the autonomic system that would normally be activated during genitally stimulated orgasm. In other words, the physiological responses were not "reflexive" responses to genital stimulation but were generated intrinsically by the brain. Orgasms during sleep are only one of the multiple contexts in which orgasm can occur apparently independent of genital sensory activity. The question of how nocturnal seminal emissions ("wet dreams") can occur independent of mechanical penile stimulation is worthy of neurophysiological research.

"Phantom" Orgasms

"Phantom limb," or "phantom limb pain," is a phenomenon in which a person who has suffered amputation of a limb feels that the limb is still present, often in severe pain. In such cases, it is, of course, the limb—not the pain—that is the "phantom." The pain is all too real.

John Money (1960) incorporated the concept of "phantom orgasm" in characterizing men and women with spinal cord injury

(SCI) and no genital sensation who experienced orgasm in their sleep. In this case, at least in the men, the "phantom" is the orgasm experienced as genital. Of fourteen men who had SCI between C5 and L1 (cervical 5 and lumbar 1), all had experienced orgasm before their injury. Eight of the men reported experiencing sexual intercourse in their dreams before injury, and five of these men reported orgasm imagery in dreams after their injury. In Money's words, these paraplegic men

> had no genitopelvic gratification (none ejaculated after their injury). It is therefore all the more remarkable a phenomenon that some of them had orgasm imagery in dreams almost as vividly as though it were the real thing . . . [This finding] offers conclusive evidence that cognitional eroticism can be a variable of sex entirely independent of genitopelvic sensation and action. The brain, in other words, can work independently of the genitalia in the generation of erotic experience, just as the genitalia of paraplegics can work reflexly and independently of the brain . . . The occurrence of orgasm imagery in the sleep dreams of paraplegics may be regarded as a special example of phantom imagery. It is of interest that this phantom experience was restricted to sleep. Awake or asleep, there were no other reported examples, from any of the patients, of phantom sensation or imagery attributable to the genitalia.

Money (1960) also described the case of a 32-year-old woman who had been injured in a fall three years earlier that produced a fracture dislocation at C6 and C7. The injury left her incontinent and paralyzed except for minimal toe movements, which disappeared following rhizotomy (surgical cutting of sensory nerve roots to reduce leg spasms). The woman stated that "when I have had a sexy dream I have always . . . reached a climax." The orgasm dreams were rare. The woman estimated that she had had six orgasm dreams in the three years since her injury.

After Spinal Cord Injury in Women

Evidence for Orgasms in These Women

There are anecdotal reports that women with complete spinal cord injury experience orgasms (Whipple, 1990). Early reports described women diagnosed with "complete" SCI who could perceive genital sensations, including orgasm (Cole, 1975; Kettl et al., 1991); these findings were later confirmed (Komisaruk & Whipple, 1994; Sipski & Alexander, 1995; Sipski, Alexander & Rosen, 1995; Whipple, Gerdes & Komisaruk, 1996; Komisaruk, Gerdes & Whipple, 1997). In Komisaruk, Whipple, and colleagues' research studies, women with complete SCI reported responding to vaginal or cervical self-stimulation—some could perceive it and some experienced orgasms in response to it. Some of these women reported that their physicians had informed them that they must be imagining the sensation, because their spinal cord had been severed. Several of the women were upset by the discrepancy between their own perception of genital sensibility and their physician's assertion of its impossibility.

In one study, Sipski, Alexander, and Rosen (1995) found that 52 percent of women with SCI experienced orgasm, that the degree and type of SCI was not related to the occurrence of orgasm, and that there were no characteristics that allowed prediction of which women with SCI would be able to experience orgasm. They did find, however, that the women who experienced orgasms had a higher sex drive and greater sexual knowledge than those who did not. In a second study, Sipksi, Alexander, and Rosen (2001) found that 59 percent of women with various levels and degrees of SCI could experience orgasm. Regarding the nature of the orgasms, independent raters were unable to differentiate between subjective descriptions of orgasm from women with SCI compared with noninjured controls. The authors have proposed that orgasm is a reflex response of the autonomic nervous system, but they did not explain how that activity could be perceived if neural transmission from the genitals to the brain was blocked by the SCI (Sipski, Alexander & Rosen, 1995; Sipski, 2001). We believe there is a different explanation for women with complete SCI being able to perceive genital sensation.

A Neurological Explanation

In cases of complete SCI, at least some genital sensation could occur if the injury extended as high as spinal cord T11 (thoracic 11). This is based on evidence that the hypogastric nerves ascend in the sympathetic chain and enter the spinal cord at T10 to T12 (Bonica, 1967; Netter, 1986; Giuliano & Julia-Guilloteau, 2006; Hoyt, 2006). Consequently, there would still be intact sensory nerve input to the spinal cord—and then access to the brain—at least via the hypogastric nerves, even if the SCI were as high as T11. The hypogastric nerves convey sensory input from the uterus and cervix, as evidenced by nerve recordings that mapped the sensory fields and zones of entry into the spinal cord of the genital sensory nerves in the female rat (Komisaruk, Adler & Hutchison, 1972; Kow & Pfaff, 1973–74; Peters, Kristal & Komisaruk, 1987; Berkley et al., 1990; Cunningham et al., 1991).

Evidence that the hypogastric nerves convey uterine sensory activity after such SCI has already been published. Berard (1989) reported that pregnant women with SCI below T12 can feel uterine contractions and movement of their fetus in utero.

To test the functionality of this hypogastric nerve pathway, Komisaruk, Whipple, and colleagues studied a group of ten women whose complete SCI was below T10 (thus presumably allowing some genitospinal input to the brain via the hypogastric nerve entry at T10) (Komisaruk & Whipple, 1994; Whipple, Gerdes & Komisaruk, 1996; Komisaruk, Gerdes & Whipple, 1997). Women could indeed feel cervical self-stimulation. They also could feel the application of a cervical stimulation probe by one of the investigators, and showed significant analgesia measured at the fingertips during cervical self-stimulation (an objective measure of response to cervical self-stimulation). Furthermore, two of the women experienced orgasms during the self-stimulation. Of greater interest, a group of six women with complete SCI at or above T10 (as high as T7) had perceptual responses comparable to the other, "lower injury" group. That is, four of the six had perceptual responses to cervical stimulation by the investigator and could feel the cervical self-stimulation, all six experienced analgesia at the fingertips, and one experienced orgasms in the laboratory. In addition, in both groups of women, all but one reported that they commonly experience menstrual discomfort.

Evidence That a Genital Sensory Pathway Goes Directly to the Brain, Bypassing the Spinal Cord

Based on these unexpected and surprising findings, we hypothesized that the women with the higher level of complete SCI experienced the vaginocervical stimulation via the vagus nerves (i.e., cranial nerve 10), which bypasses the spinal cord in its course to the brain (Komisaruk & Whipple, 1994; Whipple, Gerdes & Komisaruk, 1996; Komisaruk, Gerdes & Whipple, 1997; Whipple & Komisaruk, 1997).

The traditional view of the pathway by which genital stimulation reaches the brain is via the ventrolateral spinothalamic tract (Beric & Light, 1993). In cases of traumatic SCI, it has been reported that if this tract is interrupted, genital stimulation–induced orgasm is blocked in women and men (Beric & Light, 1993). It is curious that this pathway also contains axons that convey pain impulses to the brain. In cases of intractable pain of cancer, the spinothalamic tract may be therapeutically transected by surgery. In such cases of bilateral interruption of the spinothalamic tract, orgasm is abolished in men (Monnier, 1968). In one case of a man with intractable pain, surgical transection of the spinothalamic tract blocked genitally stimulated orgasm along with blocking the pain (Elliott, 1969). The pain blockage persisted for several months, after which the pain reappeared. Concurrently, his genital orgasmic response reappeared.

The plausibility of our hypothesis that the vagus nerves provide an additional genital sensory pathway in women is as follows. Evidence for a vaginocervical sensory role for the vagus was first presented by Guevara-Guzman and colleagues based on their studies in the laboratory rat (Ortega-Villalobos et al., 1990). They reported that the neural tracer horseradish peroxidase, when injected into the cervix, produced labeling of neurons in the nodose ganglion, which is the dorsal root (i.e., sensory) ganglion of the vagus nerve. More recently, innervation of the uterus and cervix by the vagus nerves in the rat was confirmed by Papka and colleagues (Collins et al., 1999). Support for a vaginocervical sensory role for the vagus nerve in the rat was also provided by functional studies. Vagal electrical stimulation produced analgesia in rats (Maixner & Randich, 1984; Randich & Gebhart, 1992; Ness et al., 2001) and in humans (Kirchner et al., 2001). And vaginocervical probing in rats produced analgesia even after combined bilateral transection of the known genitospinal nerves—pudendal, pelvic, and hypogastric (Cueva-Rolon et al., 1996). In the same individual rats, the analgesia was abolished after subsequent bilateral transection of the vagus nerves. In a separate study in rats, significant pupil dilation (indicative of brain involvement) in response to vaginocervical stimulation persisted, although at a diminished magnitude, after total surgical ablation of the spinal cord at the midthoracic level (T7); subsequent bilateral transection of the vagus nerves at the subdiaphragmatic level abolished the pupil-dilation response (Komisaruk, Bianca, et al., 1996). Furthermore, electrical stimulation of the central end of the transected vagus nerve (the part still connected to the brain) produced marked and immediate pupil dilation, which mimicked the effect of vaginocervical stimulation (Bianca et al., 1994). In other experiments, Hubscher and

Berkley (1994) showed that neurons of the nucleus of the solitary tract (NTS, for "nucleus tractus solitarii") in rats responded to mechanical stimulation of the vagina, cervix, uterus, or rectum, and vagotomy altered these responses. Thus, various lines of evidence—anatomical and functional—support a genital sensory role for the vagus nerves, at least in the laboratory rat.

Evidence That the Vagus Nerves Are Genital Sensory in Women

To ascertain whether the vagus nerve functions comparably in women, fMRI (functional magnetic resonance imaging) studies were used to observe whether vaginocervical self-stimulation produces activation of the region of the brain to which the sensory vagus projects—that is, the NTS, in the medulla oblongata. The study sample was restricted to women whose complete SCI above T10 was due to mechanical interruption by gunshot wound, rather than compressive injury, in order to reduce the possibility of undetected residual spinal cord pathways. The region of the NTS was identified in a prior fMRI study in which the subjects tasted a sweet-salty-sour-bitter liquid mixture to activate the superior region of the NTS (Komisaruk, Mosier, et al., 2002), which conveys gustatory sensory activity (Travers & Norgren, 1995). Activation of this region was first checked by squirting a 1-milliliter sample of the tasting mixture into the subject's mouth and recording the fMRI activation pattern. The women then performed vaginocervical self-stimulation.

In humans, the NTS is a long tubular nucleus (column of neurons) situated vertically in the medulla oblongata of the brainstem, which itself is situated vertically as an extension of the spinal cord. The NTS has been shown to have a viscerotopic organization in the rat. *Viscerotopic* implies that, if extrapolated to humans, oral input would be at the uppermost region, followed sequentially by input from esophageal, gastric, and intestinal stimulation, in descending order toward the lowermost region of the NTS (Altschuler, Rinaman & Miselis, 1992). Based on those considerations, responses to cervical self-stimulation are likely to occur at the lowermost region of the NTS—that is, at the NTS pole opposite and below that activated by the tasting mixture. The fMRI findings supported this hypothesis (Komisaruk, Whipple, Crawford, et al., 2004) and thereby provided support for the concept that the vagus nerves can convey sensory activity from the genitals in women, bypassing complete SCI at any level. Thus the "mysterious" orgasms from cervical self-stimulation in women whose spinal cord is severed are evidently conveyed by a previously unrecognized pathway—the vagus nerves.

From Electrical Stimulation of the Brain

Much of what neuroscientists have learned about the role of specific brain components in the control of behavior and physiology is based on direct electrical stimulation of the brain or administration of chemicals directly to the brain. A pioneer in the field was the neurosurgeon Wilder Penfield, of the Montreal Neurological Institute, whose work in the 1950s captured the public imagination to the degree that a street in Montreal is now named after him.

The "Little Person" in the Brain

Using a handheld electrode, Penfield stimulated the brain surface in awake patients who had been prepared for neurosurgery to remove brain regions that were triggering epileptic attacks (the temporarily displaced scalp and skull were locally anesthetized). Penfield and Rasmussen (1950) found that in a band of tissue that stretches across the cortex like the band connecting two earmuffs, there is a relatively orderly arrangement of responses to different regions of the body. That is, when the brain region is stimulated to which the patient reports toe sensation, the adjacent brain region elicits the report of leg sensation, then the next adjacent brain region elicits a report of thigh sensation, and so on. Using this method, Penfield was able to generate a sensory map of the body, termed a "homunculus" (Latin for "little person"). Just in front of that band of tissue, Penfield found a parallel map of the body, but, in this case, electrical stimulation of specific locations on the cortex provoked muscular movement of the different parts of the body. This "motor homunculus" closely paralleled the sensory homunculus.

When Penfield stimulated the part of the cortex closer to the base of the brain—in a region known as the temporal lobe, because it is a lobule of the cortex situated near the temples of the skull—he observed a different type of response in his patients. These were much more complex. They were dreamlike memories, often producing a feeling of déjà vu. Perot and Penfield (1960) called these "experiential hallucinations." They described the effect as follows:

> An experiential response [to temporal cortex stimulation] . . . may consist of a detailed experience, that is, a scene with the appropriate sounds and accompanying emotions which the patient usually recognizes as having occurred in the past; or it may be limited to a picture, a spoken word, voices, music or fragments of experience that he may or may not be able to identify . . . [They may be] (1) auditory—experiences with no apparent visual component such as words or music, (2) visual—scenes, persons or objects, (3) combined visual and auditory experiences, and (4) unclassified experiences, e.g. "a dream," "a flashback," or "a memory."

Thus, the memory, the subject's associations to that memory, and the feelings associated with the event were all evoked by the electrical stimulation.

Electrical stimulation of another brain region—the insular cortex—produced sensory effects such as "alimentary sensations" from mouth to anus, abnormal and disagreeable taste and nontaste sensations from the tongue, and gastric sensations that may have resulted from increased gastric motility (Penfield & Faulk, 1955).

Sexual Feelings

While Penfield did not report orgasmic responses in his patients (Penfield & Rasmussen, 1950; Penfield, 1958), two other neurosurgeons, using different methodologies and stimulating different brain regions, did elicit reports of orgasms by their patients. Robert Heath, at Tulane University in New Orleans, and Carl Wilhelm Sem-Jacobsen, in Norway, were the pioneers in this methodology. They implanted electrodes deep into the brain rather than on the surface, then attached them firmly to the skull, so that patients could be stimulated while up and about. These are termed "chronically implanted" electrodes.

Sem-Jacobsen (1968) described a patient with a stimulating electrode in the "posterior part of the frontal lobe, 2 cm from the midline" (the location was not more precisely described) who responded to electrical stimulation "with a nonsexual type of orgastic sensation. The patient liked it and wanted to be stimulated again, but when suddenly he became satisfied, he did not want any more electrical stimulation . . . he reported feeling 'relaxed, pleasant . . . it's like a sexual pleasure. No smell. No taste. I feel it in the whole body.' Upon realizing that this was definitely a sexual response, no further stimulation was made." Another of Sem-Jacobsen's patients enjoyed the stimulation and asked for more. This patient's responses included trembling, deep breathing, flushing, sudden relaxation, smiling, and ejaculation.

Heath (1964) described several case studies of patients who received electrical stimulation of the brain. Patient B-7 received chronically implanted electrodes preparatory for brain surgery to treat epilepsy. Heath used an interesting method in which, rather than applying the electrical stimulation, he provided patients with a wearable control panel that enabled them to apply self-stimulation at will. In one case, he provided a man with a button that controlled electrical stimulation to an electrode implanted into the septal area and another button to an electrode implanted into the midbrain tegmentum. Tegmentum stimulation produced "intense discomfort and induced fear." However, "when asked why he pressed the septal button with such (high) frequency, the patient said the feeling was 'good' and made him feel as if he were building up to a sexual orgasm. He was unable to achieve the orgastic end point, however, and explained that his frequent, sometimes frantic, pushing of the septal button was an attempt to reach a 'climax' although at times this was frustrating and produced a 'nervous feeling.' "

Three of Heath's male patients experienced penile erection during septal stimulation. Heath related the experiences of patient B-10, a man with psychomotor epilepsy, and of several patients with intractable pain:

> [Patient B-10's] response to septal stimuli frequently had sexual association. Regardless of his baseline recording or the topic under discussion, the patient would introduce a sexual subject, usually accompanied by a broad grin, upon receiving septal stimulation. When asked about his response, he said, "I don't know why that came to mind—I just happened to think of it."
>
> The pleasurable feelings from mesencephalic (midbrain) stimulation were not associated with sexual thoughts. The patient also liked the response he obtained with stimuli to the amygdaloid nucleus and the caudate nucleus, and he stimulated other septal electrodes and one other electrode in the amygdaloid nucleus a considerable number of times . . . Striking and immediate relief from intractable physical pain was consistently obtained with stimulation to the septal region of three patients with advanced carcinoma . . . The patients were stimulated at intervals ranging from two times per day to once every 3 days over periods of 3 weeks to 8 months. Stimulation to the septal region imme-

diately relieved the intense physical pain and anguish, and the patients relaxed in comfort and pleasure.

Orgasms from Electrical Brain Stimulation for Parkinson's Disease

In *Brain Control,* Elliot Valenstein (1973) describes the response of one individual with Parkinson's disease (PD) to electrical brain stimulation:

> Dr. N. P. Bechterewa of the Institute of Experimental Medicine in Leningrad . . . has studied the effects of brain stimulation in patients suffering from Parkinson's disease and other brain disorders disrupting bodily movements . . . Bechterewa reports several cases in which stimulation of the ventrolateral thalamus or adjacent regions evoked erotic and other pleasant sensations. In one case, that of a 37-year-old woman with postencephalitic Parkinsonism, stimulation evoked very pleasant sexual sensations that led to an orgasm. The patient began to visit the electrophysiology laboratory more frequently, and she initiated conversations with the assistants. She also waited for them in the corridors and the hospital garden and tried to find out when the next stimulation session was due. The patient seemed to be particularly affectionate toward the person on whom the electrical stimulation depended and she displayed dissatisfaction when her requests for additional sessions were not granted.

Quoting this passage is not to suggest brain stimulation as a therapy for parkinsonism; however, significant deleterious effects on sexuality and orgasm have been reported by men and women with PD. Waters and Smolowitz (2005) reported that of 15 men and 10 women with PD, 75 percent of the women said their frequency of orgasm was reduced since diagnosis with PD, 38 percent said they were unable to experience orgasm, and 71 percent reported a decrease in sexual interest. Among the men, 80 percent said their sexual frequency had decreased since the diagnosis of PD, 44 percent reported decreased sexual interest and drive, and 54 percent were not able to achieve an erection. Overall, a decrease in the affected partner's sexual interest was noted by 54 percent of the spouses.

Based on the extensive evidence of the role of dopamine in sexual response and orgasm, it is likely that the loss in sexual response and orgasm in PD—as well as the loss of motor function—is closely related to the deterioration of the dopaminergic neuron system, which is diagnostic of this disease.

From Chemical Stimulation of the Brain

Heath extended the method of electrical stimulation of the brain by using different neurochemicals inserted into chronically implanted cannulas in the brain (Heath & Fitzjarrell, 1984). A 33-year-old woman underwent an operation for epileptic seizures. She received cannulas into the septal region bilaterally. "With introduction of acetylcholine [one of the major neurotransmitters] directly into the septal region," the authors reported, "the patient became euphoric (often experiencing sexual orgasm) in association with continuous bursts of high-amplitude spindling [a particular type of electroencephalographic activity] focal in the septal region, activity that gradually diminished over a thirty-minute period."

Heath (1964) reported that after injection of acetylcholine into the septum of a female epileptic patient (patient B-5) when she was in a period of depression, anguish, and despair, these feelings "were supplanted within minutes by pleasurable feelings. Consistently, strong pleasure was associated with sexual feelings, and in most instances the patient experienced spontaneous orgasm . . . This patient, now married to her third husband, had never experienced orgasm before she received [this] chemical stimulation to the brain, but since then has consistently achieved climax during sexual relations." And Heath reported on another case: "The most striking sexual response occurred in a female epileptic patient with administration of acetylcholine to the septal region. A sexual motive state of varying degree consistently developed in association with the pleasure response resulting from stimulation to the septal region, whereas a specific kind of behavior was not evident in association with the pleasure response resulting from stimulation to other areas of the brain."

Subsequent ethical concerns and stringent controls on research on human subjects interrupted psychosurgery research with these invasive brain-stimulation procedures. However, certain types of brain stimulation are still used clinically, such as for controlling the tremor of PD and intractable pain in some cases.

From Electrical Stimulation of the Spine

Recent media reports described the studies of anesthesiologist and pain specialist Stuart Meloy, in Winston-Salem, North Carolina. Meloy (2006) found that when he applied electrical stimulation through the spine for the treatment of chronic back pain, 10 of 11 of his female patients, some of whom claimed they did not experience genital stimulation–induced orgasms, reported experiencing one or more orgasms during the electrical stimulation. The type of electrical stimulation Meloy used is thought to counteract pain by nonselectively stimulating many different types of nerve fibers simultaneously in the spinal cord and thus "drowning out" the pain-producing signals originating from the chronic back problem. Apparently, this electrical stimulation procedure effectively bypasses and mimics the sensory signals that would otherwise be elicited from the genitals, generating the feelings of orgasm. Without published details of his procedures and results, however, we can only make an educated guess about the basis for the reported effect.

During Epileptic Seizures

Much of what is known about how the brain produces orgasms is based on studies of epileptic seizures. There are numerous reports of men and women who describe orgasmic feelings just before the onset of an epileptic seizure. This experience has been termed an "orgasmic aura" (Calleja, Carpizo & Berciano, 1988; Reading & Will, 1997; Janszky, Szucs, et al., 2002; Janszky, Ebner, et al., 2004). The most common brain region from which these orgasmic auras originate is the right temporal lobe of the forebrain, which contains the hippocampus and the amygdala. The site of origin of the epileptic activity is ascertained by EEG (electroencephalograph). The aura may have a spontaneous onset or may be triggered by some specific stimulus—for example, orgasmic aura was triggered in a woman when she brushed her teeth (Chuang et al., 2004).

While seizure-related orgasms may be described as "unwelcome" (e.g., Reading & Will, 1997), in other cases they have been described as pleasurable. One woman was reported to have refused antiepileptic medication or brain surgery because she enjoyed her orgasmic auras and did not want to have them eliminated (Janszky, Ebner, et al., 2004).

An interesting and instructive observation in the case of orgasmic auras is that they are not necessarily experienced as involving genital sensation, but instead as "nongenital orgasms." By contrast, there are other reports of epileptic seizures that originate in the genital projection zone of the sensory cortex. Individuals report that they experience genital sensation that develops into an orgasm, and the orgasm feels as if it were generated by genital stimulation (e.g., Calleja, Carpizo & Berciano, 1988).

There are several reports of temporal lobe epilepsy that resulted in orgasmic feelings. Blumer (1970) described 29 of 50 patients, male and female, with temporal lobe epilepsy who were characterized by a global *hyposexuality:* "Most outstanding was their inability to experience orgasm." Twenty of the 29 had experienced orgasm less than once a year; 10 had experienced orgasm once or never. After these patients received temporal lobectomy, their seizures ceased and they became chronically *hypersexual.* In two patients seizures recurred, at which time their hypersexuality ceased. These patients underwent removal of the anterior portion of the temporal lobe on one side that was generating the seizure (i.e., unilateral removal of the epileptogenic temporal lobe, removing the limbic structures of the medial portion of the gyrus—presumably hippocampus and amygdala). One male patient underwent temporal lobectomy, but about one year later the seizures recurred. At that time, about 20 minutes after the attacks "he would seek sexual relations with his wife . . . His wife had started to look forward to this happening. Normally he would not seek sexual relations more than once a week. However, if a seizure occurred following sexual relations—even only one hour later—he would desire sexual relations again." Another patient rejected surgical intervention for his seizures: "At that time his wife told the neurosurgeon that he was regularly demanding intercourse immediately after his attacks. At times when he was having several attacks a day, his impatient demands were difficult for his wife, but she always acquiesced. By contrast, in the absence of seizures several weeks might pass without his experiencing sexual arousal." One of Blumer's patients "experienced the feeling of sexual climax with each of his seizures."

Fadul et al. (2005) described a "focal paraneoplastic limbic encephalitis presenting as orgasmic epilepsy" in a 57-year-old woman with a two-month history of daily spells that consisted of a sudden pleasure-provoking feeling that was described as "like an orgasm." The feeling lasted for 30 seconds to 1 minute. MRI revealed a tumor in the left anterior medial temporal lobe of the forebrain, probably due to metastasis from lung cancer that was diagnosed when the woman first sought medical care. The EEG showed a focal left midtemporal abnormality. After antitumor medication (carbamazepine) the tumor regressed and the spells subsided.

The reports that epileptic seizures can generate orgasmlike feelings suggest a basic commonality between the two phenomena. Epileptic seizures are characterized by abnormal synchronous activation of a large numbers of neurons, followed by their synchronous inactivation, then shortly thereafter by their synchronous reactivation. It is likely that the rhythmical and voluntary movement–generated timing of genital stimulation that ultimately generates orgasm also produces synchronous activation of large numbers of brain neurons, although in a more precisely regulated pattern. A consequence (and probably a function) of this regulated synchronous activity in orgasm is the activation of high-threshold systems, such as the system that controls ejaculation.

The evidence that the ejaculatory system has a relatively high threshold is that under normal conditions, rhythmical and timed genital stimulation is necessary to recruit neural elements to a higher and higher state of excitation, which climaxes with ejaculation. The ejaculation cannot be elicited at lower levels of excitation, and thus it is characterized as a high-threshold system. In the case of ejaculatory orgasm, the genital stimulation is channeled into relatively specific and coordinated systems, such as coordinating vigorous thrusting movements with the ejaculatory event. By contrast, in an epileptic seizure, the mass synchronous neural activation becomes abnormally diffuse and can "spill over" into motor systems that are not normally activated simultaneously, resulting in uncoordinated limb movements, loss of balance, uncoordinated facial and tongue movements, and so forth, as in a grand mal seizure.

Thus, the mass neuronal activation that characterizes an epileptic seizure bears a resemblance to the mass neuronal activation that characterizes orgasm. It is perhaps this similarity that can generate the orgasmlike feelings during epileptic seizures.

Multiple Orgasm

Many books are available that profess to describe ways of inducing multiple orgasm in men and women—for example, *Any Man Can* (W. Hartman & Fithian, 1984); *Orgasm: New Dimensions* (Kothari, 1989); *The Multi-orgasmic Man* (Chia & Arava, 1996); *The Multi-orgasmic Couple* (Chia et al., 2000); and *The Multi-orgasmic Woman* (Chia & Abrams, 2005). We know of only four published studies on multiple orgasm in men, and slightly more on multiple orgasm in women. It seems that multiple orgasm is recognized in some women but not as easily recognized in men. Among the classic studies, Kinsey and colleagues (Kinsey, Pomeroy & Martin, 1948; Kinsey et al., 1953) reported multiple orgasms in both men and women, whereas Masters and Johnson (1966), based on their laboratory observations, reported multiple orgasm in women but not in men.

There is no clear definition of multiple orgasm. Successive orgasms in women may occur within a few seconds or a minute or two (Kinsey et al., 1953). Masters and Johnson (1966) reported that the state of arousal between multiple orgasms does not fall below what they termed the "plateau phase" level. Hite (1976) suggested that if there are pauses between orgasms that require restimulation, then the term *sequential* rather than *multiple* orgasm should be used; multiple orgasms involve continuous stimulation with no interruption.

Kothari (1989) defined *multiorgasm* as "a function of sustained arousal after each orgasmic episode that culminates again in orgasmic intensity by further stimulation." Men and women, he continued, "have similar underlying explanations with respect to the psychodynamics of multiorgasm. For the male, a cognitive orientation to experience multiorgasm is essential along with strong pubococcygeus muscle . . . A woman's multiorgasmic potential is the outcome of her physioanatomical mould. For the same reason, a considerable number of women experience multiorgasm by adequate continuous stimulation and they need less cognitive readiness than males." Unfortunately, Kothari does not specify the essential ways in which he considers women's "physioanatomical mould" to differ from that of men.

W. Hartman and Fithian (1985) pointed out that as early as 2968 BC, in China, there were writings that described male multiple orgasms, as well as mention of multiple orgasm in Taoism, Tantra, Vishrati (in ancient India), Imsak (in the Arab world), Chira in the *Kama-sutra,* and in more modern references.

Three published studies on multiple orgasms in males focused on multiple orgasm without ejaculations, except in the last orgasm of the sequence (Robbins & Jensen, 1978; M. E. Dunn & Trost, 1989; Kothari, 1989). In all of the studies, the participants employed techniques to delay the occurrence of ejaculation.

A man reported that he had a multiple-orgasm response pattern with uninhibited ejaculation at each orgasm and that he had had this experience since the age of 15, when he first ejaculated. In the laboratory, he experienced six ejaculatory orgasms from self-stimulation within 36 minutes. He maintained an erection during the entire period. There were significant increases in all physiological parameters measured during the orgasmic conditions. Blood pressure, heart rate, pupil diameter, and reported level of arousal all increased significantly over resting conditions during each orgasm (Whipple, Myers & Komisaruk, 1998).

The essential physiological differences between the single and multiorgasmic pattern are not known. It may be that multiple orgasm is a phenomenon that some men and women experience spontaneously and others are taught or teach themselves. Multiple orgasm is not necessarily a goal one should strive for; it is another form of orgasmic experience that some people enjoy.

After Sex-Change Surgery

Male-to-Female Surgery

Knowledge of the sensory fields of the genital sensory nerves is utilized in reconstructive genital surgery and sex-change surgery. In one type of male-to-female sex-change surgery, the phallic skin and its sensory pudendal nerve and vascular supply are retained and form the inner lining of an artificial "vaginal tube"; the tube is sutured closed at the long end opposite the "vaginal" opening, to form a test tube–shaped structure. An incision is made in the perineal skin between the anus and the urethra, and a cavity is created, pushing up toward, but not into, the abdominal cavity. The vaginal tube is inserted into the cavity, and the open end is sutured to the perineal skin. The glans penis, along with its sensory nerves, is shaped into a clitoris, and the labia are formed from the scrotum (Krege et al., 2001; personal communication, 2006). Orgasm is retained in response to mechanical stimulation of this tissue during sexual intercourse. This is not surprising, because stimulation is applied to the original (penile) skin with its same nerve supply. Also, indirect stimulation of the prostate through the "neovaginal" wall could add pleasurable sensations and orgasm.

However, less intuitively obvious is the report of orgasm after the reconstructive surgical procedure that creates an artificial vagina from a segment of *intestine,* the rectosigmoid colon. This surgical procedure has been used in male-to-female operations after penectomy and orchiectomy (surgical removal of the penis and testicles), in women with congenital vaginal atresia (absence of the vaginal opening), and in women with short vaginal length after cervical cancer surgery (S. K. Kim, Park, et al., 2003). An incision is made just in front of the anus into the abdominal cavity. A segment of the rectosigmoid colon is removed, keeping its blood supply and nerve supply (including the hypogastric nerve) intact; the inner end is sealed and the outer end is sutured to the perineum (the region of skin between the anus and the scrotum). The authors reported that "twenty-four of the 27 patients (89 percent) experienced orgasm during intercourse; 10 of the 24 (42 percent) had male-type orgasm with ejaculation and the other 14 (58 percent) had orgasm without ejaculation. The male-type orgasm pattern of the transsexuals changed to the female-type orgasm pattern gradually." In response to an inquiry to the authors of the study, Dr. S. K. Kim (personal communication, 2006) explained that in the case of male-to-female surgery, orgasm results from stimulation of the reduced amount of penile tissue that is surgically modified to produce a "neoclitoris." In addition, because "the rectosigmoid flap is a sensory flap, it contributes pleasurable sensations." Ejaculation results from stimulating the neoclitoris, because the prostate and its ducts remain intact. The ejaculation response "disappeared slowly . . . after several years." In the case of women who undergo surgery, orgasm occurs in response to stimulation not of the rectosigmoid flap but of the clitoris.

A similar surgical procedure was reported by Jarolim (2000). When it was not possible to use the patient's penile skin to create the "vagina," the surgeon used an intestinal segment (rectosigmoid colon). "Some pleasure sensation and vibration was possible in this segment," Jarolim reported, "because of the autonomic innervation which travels along the vessels [referring to the blood vessels that were left attached to the tissue] . . . surgical gender reassignment of male transsexuals resulted in replicas of female genitalia which enabled coitus with orgasm." The orgasms result from stimulation of the initially penile skin.

Female-to-Male Surgery

Female-to-male surgery is less common. In this procedure, skin flaps are cut from the inguinal region, and tissues from the clitoris and labia minora, with their nerve and vascular supply, are added to the flaps to form a "neophallus" around a urethra. In some cases this enables voiding of urine while standing. Tissue from the labia majora is used to form a "neoscrotum." Although erection is not possible, orgasm can be elicited from stimulation of the original clitoral tissue (Jarolim, 2000).

Nongenital Orgasms

Phantom Limb Orgasm

The case of a man who described orgasms in his amputated phantom foot was described by Ramachandran and Blakeslee (1999). They reported the following conversation:

> [PATIENT:] "Doctor, every time I have sexual intercourse, I experience sensations in my phantom foot. How do you explain that? My doctor said it doesn't make sense."
>
> [RAMACHANDRAN:] "Look," I said. "One possibility is that the genitals are right next to the foot in the body's brain maps. Don't worry about it." He laughed nervously.
>
> [PATIENT:] "All that's fine, doctor. But you still don't understand. You see, I actually experience my orgasm in my foot. And therefore it's much bigger than it used to be because it's no longer just confined to my genitals."

Maps of the sensory cortex show that sensation from the foot projects to the cortex immediately adjacent to the region receiving sensation from the genitals. It is likely that after amputation of the foot, fibers of the neurons in the genital sensory cortex "invade" or "sprout" into the adjacent region, vacated by the neuron fibers that originally came from the foot. This neural reorganization is similar to another phenomenon reported by Ramachandran and Blakeslee in which a man with an amputated hand felt his phantom hand when his face was touched. The hand and face sensory regions are immediately adjacent to each other in the sensory cortex.

Orgasm of Specific (Nongenital) Body Parts

A different type of "nongenital orgasm" was described by a young male colleague of Komisaruk and Whipple (1998). He experienced these orgasms under the influence of marijuana during self-stimulation of different parts of the body.

> *Nose Orgasm.* Stimulation was applied using an electric vibrator held in place against the tip of the nose. A buildup of intensity of irritating sensation was described. The imagery evoked was described as starting as a small point of light then approaching closer and closer, getting brighter and larger, as if flying directly into the face. At the moment when the irritating sensation was unbearably intense, which was just before "collision" of the bright light against the face, a sneeze occurred, "blowing away" the light.
>
> *Knee Orgasm.* With the vibrator stimulating the knee, the quadriceps (extensor) muscle of the thigh increased in tension, while simultaneously, the image observed was of an increasingly immense panoramic scene of thousands of troops and artillery. At the reported orgasmic moment, the leg gave an extensor kick, every single element in the panorama made a simultaneous forward move, and a simultaneous forceful grunt was emitted.
>
> *Penile Orgasm.* When the vibrator was applied to the tip of the penis, an image of an ocean liner appeared in the distance, being raised from the depths of the ocean by an uplifted hand. A great effort, perceived as a growing tension in the postural muscles of the trunk and limbs, was mobilized in which the hand was raising the increasingly massive ocean liner. The ocean liner then burst forth into the sunlight in a fountain of spray at which moment, peak muscle contraction, actual

> ejaculation, and laughter all erupted simultaneously. This last orgasm was a genital orgasm, but stated in the same context as the nose and knee orgasms. It had the same qualities of imagery, and muscle tension appropriate to the imagery, as did the other types of orgasms, except it was in the genital system and included the visceromotor response of ejaculation. Stated alternatively, it was similar in form to the non-genital orgasms, but it was a genital orgasm. This description indicates that there was coherence among somatic, visceral, and cognitive activity leading up to the orgasmic moment.

The effects of the marijuana may have been to break down the inhibitory pathways that normally separate waking from dream imagery, thereby revealing associations that may otherwise occur only in dreams, hallucinations, synesthesia (the [con]fusing of two different senses, e.g., "tasting shapes" [Cytowic, 1998]), psychosis, and other altered states of consciousness. The orgasms, while markedly different from each other, were all described as manifesting dreamlike imagery that was related in an understandable way to the function of the specific body part that was stimulated and expressed the orgasm. In each of the three types of orgasms described above, each component—skeletal motor, respiratory, and cognitive—while unique in its own "currency," was coherent with each of the others. Invariably, explosive respiratory activity (sneeze, grunt, laugh) accompanied each orgasm. Each orgasm built in a coherent, comprehensible (though dreamlike) crescendo of excitation, culminating in a synchronous climax that was described as pleasurable.

Apparently, just as pain is not restricted to any one part of the body, neither is pleasure. A characteristic of orgasmic pleasure, the perception of the body's explosive muscular expression, can be perceived not only in the genital system but also in the respiratory system and other body systems (Komisaruk & Whipple, 1998). Thus, it seems that while the genital system is particularly well-organized to mediate the orgasmic process, other body systems evidently manifest at least some of the same properties and, consequently, under appropriate stimulus conditions and sensitization may exhibit comparable activity.

The mental state that occurs during orgasm has been described as an "altered state of consciousness" (Davidson, 1980) that may lead to a state of tranquility and deep unconsciousness, to which the French have attributed the name "le petit mort," the little death. Despite anecdotal reports and common jokes about people (especially men) dropping off to sleep shortly after sexual intercourse and orgasm, the sleepiness-inducing effect of orgasm, and in some women and men the opposite effect of orgasm—hyperalertness—a literature search revealed only one laboratory research study that had actually addressed the issue. Suzanne Brissette and colleagues reported in 1985 in the journal *Biological Psychiatry* that, in a study of five men and five women in which researchers compared three conditions—masturbation to orgasm with a latency of about 15 minutes, masturbation for 15 minutes while intentionally refraining from orgasm, and reading a newspaper for 15 minutes—there was no significant difference in latency to fall asleep, duration of sleep, or duration of various stages of sleep among the conditions. Many disruptive factors could account for the lack of difference in their study—e.g., the laboratory setting, the fact that an investigator entered the room after every 15-minute period in order to remove an anal probe that was used to obtain objective evidence of orgasmic muscular contractions, the presence of EEG electrodes to measure sleep, the possible sleep-inducing effect of reading a newspaper, the combining of the data for the men and the women, and the absence of the personal physical and emotional interaction factors inherent in sexual intercourse. Thus, in the present state of knowledge of any connection between orgasm and sleep, anecdotal reports appear to characterize reality better than the existing scientific research.

The altered state of consciousness that may occur during orgasm has characteristics that are similar to epileptic aura and seizure. Indeed, as described above, there are reports of men and women with epilepsy experiencing feelings of orgasm during epileptic attacks. It is of interest that the part of the brain involved in epileptic attacks may include the specifically genital sensory projection sites, but more often it does *not* involve these sites. When it does, individuals report genital sensations during orgasm, but when it does not, individuals say they have had orgasms but no particular genital sensations. Evidently, *nongenital orgasm* is not an oxymoron.

20

The Genital-Brain Connection

Specific parts of the genitals send their sensory signals to the brain via specific nerves. This helps to explain the different quality of orgasms that result from clitoral, vaginal, or cervical stimulation and the different quality of sensations produced by penile, testicular, or anal stimulation. In this chapter, we describe the division of labor among the different genital sensory nerves and the sensations experienced when they are stimulated.

Sensations Conveyed by the Genital Nerves

The division of labor among the genital sensory nerves in females is generally considered to be among the hypogastric, pelvic, and pudendal nerves. The *hypogastric nerves* convey sensory activity from the uterus and cervix in women (Bonica, 1967; Netter, 1986; Giuliano & Julia-Guilloteau, 2006; Hoyt, 2006) and in rats (Peters, Kristal & Komisaruk, 1987; Berkley et al., 1990). The *pelvic nerves* convey sensory activity from the cervix and vagina in women (Netter, 1986) and, as well as from the midline perigenital skin, in rats (Komisaruk, Adler & Hutchison, 1972; Peters, Kristal & Komisaruk, 1987; Berkley et al., 1990). And the *pudendal nerves* convey sensory activity from the clitoris in women (Netter, 1986; Giuliano & Julia-Guilloteau, 2006) and, as well as from the perigenital skin, in rats (Peters, Kristal & Komisaruk, 1987). The pudendal and pelvic nerves enter the spinal cord at the upper sacral level, S2 to S4 (Netter, 1986) and lower lumbar level (Ding et al., 1999). The hypogastric nerves travel in the sympathetic chain and enter the spinal cord at the thoracic level, T10 to T12 (Bonica, 1967; Netter, 1986; Giuliano & Julia-Guilloteau, 2006; Hoyt, 2006). (The sympathetic chain is a series of autonomic ganglia resembling two strings of pearls, each lying alongside the vertebral column, outside the spinal cord.) There are minor species differences among humans, monkeys, cats, and rats in terms of the precise location in the spinal cord where these nerves enter, but the basic pattern is very similar.

More recent research suggests that a fourth pair of nerves—the *vagus nerves*—also conveys sensory activity from the cervix and uterus in the laboratory rat (Ortega-Villalobos et al., 1990; Collins et al., 1999) and the cervix in women (Komisaruk, Whipple, Crawford, et al., 2004).

In males, the pudendal nerves convey sensory activity from the penile skin and scrotum, and the hypogastric nerves convey sensory activity from the testes. Injury to the pudendal and cavernous nerves leading to symptoms of penile numbness or hypesthesia (lowered sensitivity) and erectile dysfunction was reported in 13 to 22 percent of men participating in a 540-kilometer bicycle race, some symptoms persisting for up to eight months (Andersen & Bovim, 1997).

Testicular pain sense—a "severe aching and nauseating pain . . . provoked when the normal testis is squeezed"—is conveyed by the hypogastric nerves. The evidence for this is that there is *no* dimi-

Simplified schema of the nerves that convey sensation from the female genital system. The sensory field is the portion of the genital system that is supplied by the nerves shown connected to it by lines. For example, sensation from the cervix is conveyed via the pelvic, hypogastric, and vagus nerves. And the pelvic nerves convey sensation from the perigenital skin (which includes the perineal skin), vagina, and cervix.

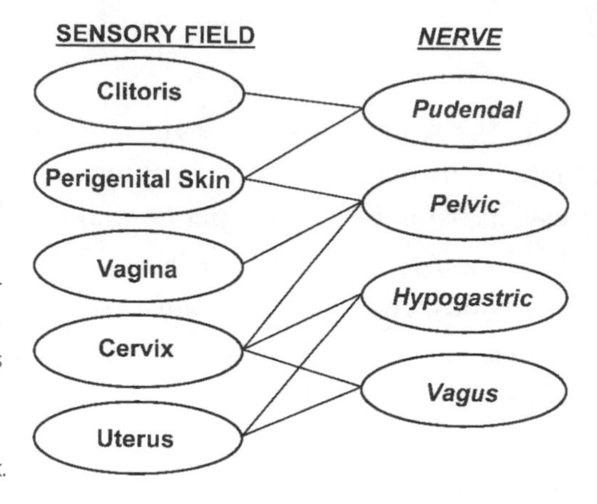

nution in testicular pain sense after complete injury of the spinal cord at the sacral level, in the conus medullaris (the end region of the spinal cord) (Hyndman & Wolkin, 1943), injury that would prevent sensory activity from entering the spinal cord via the pelvic and pudendal nerves. The hypogastric nerves enter the spinal cord at T10 to T12, passing through the sympathetic chain and consequently bypassing the sacral level of the spinal cord, as in women.

In the spinal cord, testicular pain sense is conveyed to the brain via the spinothalamic tract, which originates in the spinal cord and projects to the thalamus, which in turn projects to the cerebral cortex. If this tract is cut surgically, the pain is abolished. This surgical procedure, termed "chordotomy," also affects orgasm. From their experience with spinal cord surgery, Hyndman and Wolkin (1943) claimed that "based on 5 men and 4 women, bilateral anterior chordotomy is almost certain to abolish erection and orgasm in the male and orgasm in the female. The desire for intercourse is not abolished, but the ability of the male to carry out the sexual act is most likely to be lost." More precisely, in men, "it appears that the tracts which subserve the function of psychologically controlled erection and the sensation which accompanies orgasm do not reside in the anterior column, but are probably located close to the gray matter in the anterolateral column [nerve fiber pathways in the spinal cord that convey impulses between the genitals and brain] . . . In women, menstruation was not impaired after any type of anterior chordotomy" (an anterior chordotomy cuts the nerve fiber pathways in the spinal cord closest to the forward-facing part of the body).

Based on the findings in a patient (case 2) who had a chordotomy restricted to the more lateral regions of the spinal cord, Hyndman and Wolkin (1943) concluded that "it also appears, with special reference to case 2, that the sensation which accompanies orgasm is not a modality of pain or temperature sense, and certainly not a modality of touch sense. It is probably a specialized sensation, mediated by fibers localized deep toward the center of the cord in the anterolateral column . . . The sensation of orgasm . . . appears to be a special sense, not mediated by the spinothalamic tract."

Locating the Most Sensitive Genital Areas

While "erogenous zones" are highly individualistic, the relative sensitivity of different genital regions is more generalizable. Perry and Whipple (1981) found that the anterior wall of the vagina—the region near the 12 o'clock position, with the woman lying on her back—was most sensitive to mechanical stimulation. They found that stimulating deep-lying tissue that surrounds the urethra by pressing against the internal vaginal surface with a "come here" motion was most likely to produce orgasm. They named the region the "G spot," after gynecologist Ernst Gräfenberg, M.D., who described a "distinct erotogenic zone" in women in the anterior vaginal wall along the urethra that swells on sexual stimulation (Gräfenberg, 1950). By contrast, stimulation of the posterior wall, near the 6 o'clock position, was least likely to elicit orgasm. In a separate set of experiments on suppression of pain, mechanical stimulation of the anterior wall of the vagina was more effective in suppressing experimentally induced pain (measured at the fingers) than was stimulation of the posterior wall (Whipple & Komisaruk, 1985).

In a publication describing procedures used to perform a study of sensitivity of vaginal regions in women, Heli Alzate, professor of sexology at Caldas University School of Medicine in Manizales, Colombia, made the following delicate comment: "The paid subjects were recruited in Manizales through the good offices of a madam with whom one of the authors (HA) is acquainted." Alzate and Londono (1984) described their findings as follows:

> Among 48 women, 94 percent reported vaginal erotic sensitivity. Among 30 women tested who experienced orgasm or came close to orgasm, 73 percent showed maximal response to stimulation of the upper half of the anterior vaginal wall, 27 percent had maximal response to stimulation of the lower half of the anterior vaginal wall. In 30 percent another zone whose stimulation could elicit an orgasmic response was in the lower half of the posterior vaginal wall (some subjects showed more than one zone of maximal response). Very few reported a pleasant sensation on the cervix or posterior vaginal fornix.

In contrast, in a study of more than 20 women with or without complete spinal cord injury, the women used a specially designed stimulator to apply mechanical pressure directly to the cervix via a stimulator rod attached to a diaphragm fitted to the cervix (Komisaruk, Gerdes & Whipple, 1997). Most of the women reported feeling the stimulus. The investigators were able to measure the women's threshold of sensitivity to the pressure. Some of the women with complete spinal cord injury could feel the stimulation of the cervix, although they were less sensitive to the stimulation than noninjured women. Two of the women (without spinal cord injury) stated that when they pulled the stimulator outward, away from the cervix, the suction generated by the diaphragm against the cervix felt unusual and pleasurable.

Consistent with this description of pleasurable response to cervical stimulation, Cutler et al. (2000) found that 35 percent of 128 healthy women reported experiencing orgasm from penile stimulation of the cervix during sexual intercourse. MRI (magnetic resonance imaging) images of penile contact with the cervix are shown in the study by Faix et al. (2002). Alzate, Useche, and Villegas (1989) reported that in response to digital stimulation of the vagina by the examiner, 8 of 8 women experienced orgasm from stimulation of the anterior wall, whereas only 2 of the 8 experienced orgasm from stimulation of the posterior wall. In Cutler et al.'s study (2000), 63 percent of the women said they experienced orgasm from vaginal stimulation, and 94 percent said they experienced orgasm from clitoral stimulation. The clitoris has been characterized as the "most densely innervated part of the human body" (Crouch et al., 2004).

Using a different method of measurement—mild electrical stimulation at different intensities—Weijmar Schultz et al. (1989) tested the sensitivity of different regions of the vagina. They designated the most anterior region of the vagina (the region closest to the clitoris) 12 o'clock and the region closest to the anus 6 o'clock. They found that the 12 o'clock region of the vagina was the most sensitive. All the internal vaginal areas around the "clock" were less sensitive than the clitoris. The clitoris had slightly greater sensitivity than the labia minora. The labia majora were slightly more sensitive than the clitoris and labia minora. The abdominal skin and back of the hand were more sensitive than all these genital regions. A particular type of sensory receptor, Pacinian corpuscles, which are specialized to respond to pressure and vibration, are present in larger numbers in the clitoris and prepuce than in the labia minora (Krantz, 1958).

The Basis for Regional Differences in Vaginal Sensitivity

Another type of nerve ending, the Merkel receptors, which are specialized for responding to steady pressure, is found in the vaginal introitus region (the vaginal region near the entrance) but not in other regions of the vagina (Hilliges et al., 1995). In that region,

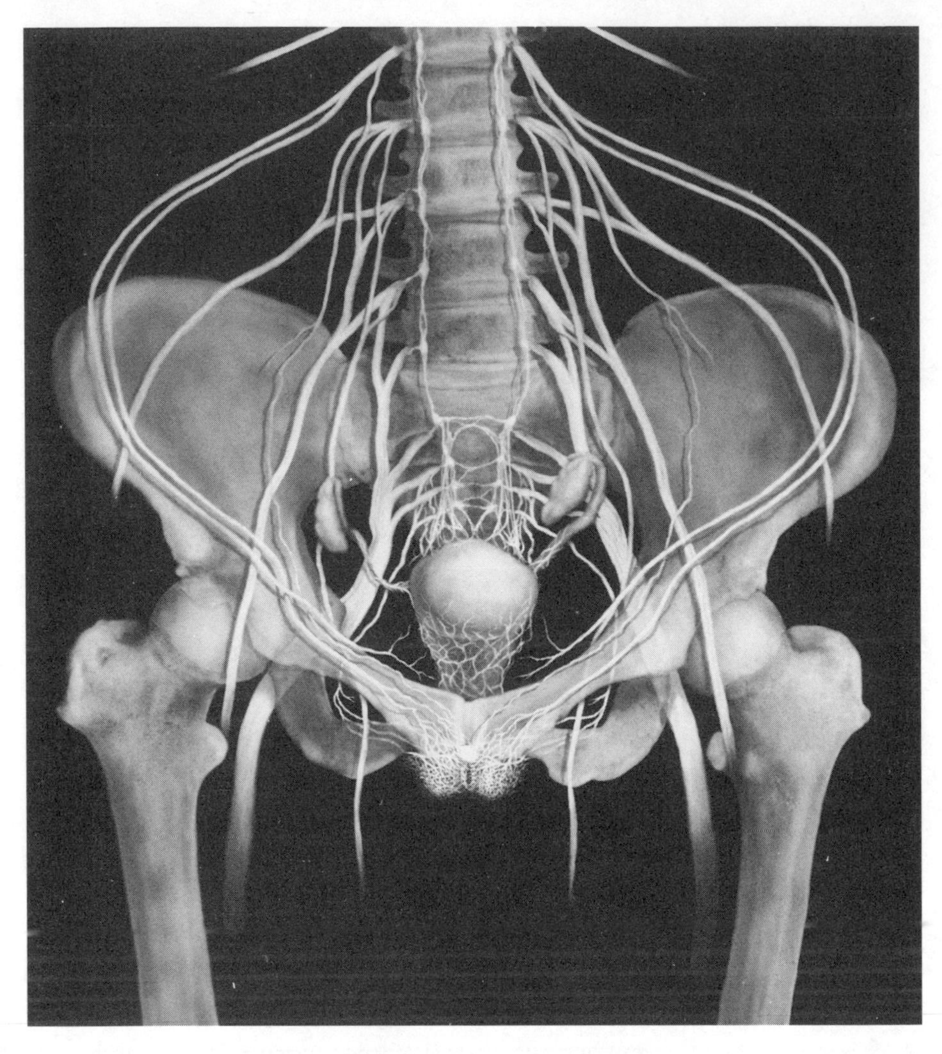

The nerve supply of the female genital system. Note the dense nerve supply of the labia, vagina, and clitoris. The nerves are especially concentrated to form the white disc at the head of the clitoris. It has been claimed that the clitoris has the densest nerve supply in the body. The paired pudendal nerves supply the clitoris and labia, and are seen coursing along the front surface of the pelvis. The paired pelvic nerves supply the vagina and course parallel to the pudendal nerves. The paired hypogastric nerves form a netlike structure between the uterus and the spine. The paired vagus nerves are not shown. (Image courtesy of Anatomical Travelogue)

researchers found a clear difference in innervation of the deep (submucosal) tissue layer of the vagina. The connective tissue of the anterior wall is richly innervated, by contrast with the posterior wall, which is more sparsely innervated.

Krantz (1958) described an unusual wavy pattern of nerve fibers in the vaginal adventitia (the connective tissue covering of the vagina in the abdominal cavity) and in the vaginal musculature. He suggested that the wavy pattern could provide a means by which the nerve endings adapt to the extreme distortion during labor and delivery. He extended his speculation further by stating, "It is conceivable that should the pattern become distorted sufficiently, one may have a partial explanation for some of the bizarre sexual patterns seen in some human beings." We wonder what he had in mind.

The sensitivity of the genitals is modified by ovarian hormones. As was first shown by Salmon and Geist (1943), systemic administration of testosterone to women whose ovaries and adrenals had been surgically removed increased the sensitivity of the skin surrounding the vagina. Other research found that estrogen administration to the laboratory rat increased the area of skin surrounding the clitoris and vagina where gentle mechanical stimulation could elicit responses recorded from the pudendal nerve, which conveys those impulses (Komisaruk, Adler & Hutchison, 1972). In other words, estrogen increased the "sensory field" of the pudendal nerve. Others found that estrogen treatment increased the sensitivity of this sensory field (Kow & Pfaff, 1973–74). This is a rapid, hormonally based change: the change in size and sensitivity of the sensory field was found to occur within just two days of the four-day ovarian cycle of the laboratory rat (Adler, Davis & Komisaruk, 1977). Subsequently, again in the laboratory rat, the sensitivity of the vaginal sensory field was shown to vary with the estrous cycle (Bradshaw & Berkley, 2000).

The Hysterectomy Controversy

There is significant and continuing controversy regarding the effect of hysterectomy on sexual response. A recent study in the United Kingdom was based on a questionnaire sent to women five years after they had been treated surgically for "dysfunctional uterine bleeding." Of the more than 8,900 women who responded to questions relating to their psychosexual function, more women who had total hysterectomy reported having "bothersome psychosexual function [i.e., loss of interest in sex, difficulty in becoming sexually excited, and vaginal dryness] than did the women who had a less invasive operation [e.g., transcervical endometrial resection/ablation or subtotal hysterectomy with and without bilateral oophorectomy (removal of the ovaries)]. Hormone therapy, although related to surgical method, did not reduce this long-term detrimental effect. The odds [of experiencing a detrimental effect of hysterectomy] were particularly high amongst women with concurrent bilateral oophorectomy" (McPherson et al., 2005).

In another recent study, measurement of vaginal and clitoral sensitivity three months after, compared with immediately before, hysterectomy showed a small but significant reduction in sensitivity to cold and warm stimuli at the anterior and posterior vaginal wall, whereas clitoral sensitivity was not affected. Three of 22 women tested reported a reduction in libido (Lowenstein et al., 2005).

By contrast, a questionnaire study of 400 women within three years of hysterectomy of various types, including total abdominal (removal of uterus and cervix) and supracervical (removal of uterus, sparing the cervix), concluded that "responses that pertained to libido, sexual activity, or feelings of femininity did not reveal significant changes" from prehysterectomy levels (Roussis et al., 2004). A similar conclusion was reached by Farrell and Kieser (2000), who reviewed eighteen reports in the literature. They concluded that while the "outcome measures were usually not validated and most studies did not consider important confounding factors . . . Most studies in this review showed either no change or an enhancement of sexuality in women who had a hysterectomy."

Part of the controversy is probably based on the great variability of critical factors, such as the conditions preceding the surgery (e.g., benign or malignant tissue), the presenting symptoms (e.g., pain), the surgical method (including the degree of nerve sparing), the extent of the surgery (whether the cervix and/or ovaries are removed in combination with the uterus), the psychological state of the woman pre- and postsurgery (e.g., whether depressed), the woman's physiological status (pre- or postmenopausal), and the nature of the data collected (type of questionnaires administered).

A recent extensive review of the literature concluded that "the research on the effect of hysterectomy that has been performed to date is not conclusive" (Maas, Weijengorg & ter Kuile, 2003). These authors pointed out that while a majority of women report an improvement of sexual functioning after hysterectomy, this may result from the relief of symptoms (such as vaginal bleeding and dyspareunia [pain during sexual intercourse]) from the diseased uterus. They also pointed out that "a minority of women reported development of sexual dysfunctions as a result of hysterectomy" and concluded that much more research is needed to clarify the issue of the effect of hysterectomy on sexual response—a position with which we agree.

Vaginocervical Inhibitory Reflex

Komisaruk and Larsson (1971) identified a powerful inhibitory genital reflex in female rats—a response to vaginal or cervical mechanostimulation. When a probe, such as a plunger from a 1-milliliter syringe, is inserted into the vagina and pressed gently against the cervix, with a force equivalent to as little as 50 to 100 grams, the rats become immobilized, with their legs rigidly extended and their abdominal muscles tensed. The immobilization is so strong that the rats can be slid, stiff-legged, across the table, and they do not walk away from the probe. And the immobilization is so potent that the rats can be turned upside down onto their backs during the stimulation, and they do not right themselves for 30 seconds or more (normally they would right themselves in a fraction of a second). If the strength of the immobilization is challenged by pinching a hind paw or the tail with toothed forceps, the normally immediate leg withdrawal response is completely blocked.

The immobilization extends throughout the body, although it is most pronounced at the hind legs and tail. The front leg withdrawal response to front paw pinch is strongly inhibited, but not totally blocked, by the vaginocervical stimulation. The eyeblink response to a thread touched gently to the cornea is strongly attenuated, if not completely blocked, as is the facial twitch response to pinching an ear. This is a true inhibitory spinal reflex. If the spinal cord is completely interrupted surgically at the midthoracic level, the rat loses the ability to locomote with its hind legs but still shows reflex withdrawal of the hind legs when the hind paw is pinched. However, vaginocervical stimulation completely blocks the leg withdrawal response to the hind paw pinch. This demonstrates that the inhibitory response to the vaginocervical stimulation is intrinsic to the spinal cord and does not *require* mediation by the brain. However, the response to vaginocervical stimulation does also involve the brain, because the cranial nerve reflexes, such as eyeblink and the facial twitch response to ear pinch, are also attenuated by the stimulation (Komisaruk & Larsson, 1971).

We believe that the leg extension–immobilization response to vaginocervical stimulation may be a normal response by the female rat during mating and parturition. During mating, the female rat stands rigidly still, extending the hind legs, elevating the rump, diverting the tail, and exposing the vaginal opening while the male mounts and intromits the penis. This further stimulates the intensity of rump elevation by the female. This mating posture (lordosis) is repeated, interspersed with episodes of the female darting away and stopping again, allowing the male to mount again, approximately eight times within several minutes, until the male ejaculates. Thus, the leg extension and immobilization are normal elements of the mating posture.

The leg extension, immobilization, and abdominal muscle tensing also occur during parturition, which is the other occasion when the female rat receives vaginocervical mechanical stimulation naturally. In addition, cervical stimulation in the rat produces a relaxation of the circumvaginal muscles, further facilitating expulsion of the fetuses (Martinez-Gomez et al., 1992). The leg extension seems to mechanically facilitate the emergence of the pups from the vagina. The abdominal muscle tensing enables parturition to occur; this is evidenced by the finding that blockage of vaginocervical stimulation–induced muscle contraction by cutting the pelvic nerves, which comprise the main vaginocervical sensory pathway (e.g., Komisaruk, Adler & Hutchison, 1972), prevents delivery of the pups (Higuchi et al., 1987).

A similar inhibitory effect occurs in women. In laboratory studies, vaginal self-stimulation inhibits leg withdrawal to toe pinch (Komisaruk, Whipple, and colleagues, unpublished observations).

Orgasm Is Not a Reflex

We believe that the potent effect of the vaginocervical stimulation in extending throughout the body is indicative of its "hard-wired" access to widespread populations of neurons. Genital stimulation can thus recruit these neuron populations and intensely stimulate the many voluntary and involuntary muscle groups that participate in the orgasmic response. The rhythmical, repetitive genital stimulation that characterizes sexual intercourse recruits the large populations of neurons in the brain into greater and greater levels of activity, generating the climactic characteristics of orgasm.

It should be emphasized that although the contractions of involuntary (smooth) muscle (e.g., in ejaculation or uterine contractions) and voluntary (striated) muscle (e.g., of the pelvic floor) that occur during orgasm can be elicited reflexively when the spinal cord is separated from the brain, it does *not* follow that orgasm is a reflex—contrary to what some have claimed. It would seem self-evident that orgasm is a *perception*, a function of activity occurring in the brain. The sensory activity generated by genital stimulation reflexively activates orgasmic muscular contractions, and the sensory activity generated by the muscular contractions creates a "positive feedback"—a cyclically building cascade of sensory stimulation that adds to the pleasurable qualities of orgasm. The muscular events by themselves do not generate orgasm unless and until the sensory activity they generate gains access to and is perceived by the brain.

The Contribution of Oxytocin to Orgasm

Although the literature is not clear on whether injection of the neurohormone oxytocin intensifies orgasm in women, sexual receptivity is intensified by administration of oxytocin in female rats (Moody et al., 1994). In rats, mechanical stimulation of the vagina and cervix induces sexual receptivity in otherwise unreceptive females (Rodriguez-Sierra, Crowley & Komisaruk, 1975). Since oxytocin stimulates vaginal, cervical, and uterine contractions by stimulating the smooth muscle of these organs, it is possible that the sensory stimulation emanating from these organs as a consequence of their muscle contraction mimics the receptivity-stimulating effect of direct mechanical stimulation. Researchers found that cutting the nerves that convey sensory activity from the vagina, cervix, and uterus (i.e., the pelvic and hypogastric nerves) abolished the receptivity-stimulating effect of injected oxytocin. This finding supports the hypothesis that, rather than acting directly on the brain or spinal cord, oxytocin intensifies sexual receptivity by stimulating the contraction of vaginal, cervical, and uterine smooth muscles, thereby generating sensory stimulation (Moody et al., 1994). There are inconsistent reports as to whether injection of oxytocin increases the perceived intensity of orgasm in women. It is possible that increased sensory activity resulting from oxytocin-induced contraction of these genital smooth muscles may generate contractions of voluntary muscles of the pelvic floor, much as vaginocervical stimulation induces contractions of abdominal muscles. In this case, the contraction of pelvic floor muscles, as pointed out by Messe and Geer (1985), could contribute to the feeling of pleasure during intercourse.

The basic concept in this model of oxytocin action is that of "reafference." That is, the initial vaginal sensory stimulation ("afference") leads to intensified activity, such as by releasing oxytocin that intensifies smooth muscle contraction in the genital tract or voluntary muscle contraction, or both, which in turn generates increased sensory input from these muscles ("reafference"); the intensified afferent activity then intensifies the pleasurable feeling of orgasm.

21

Orgasms after Brain Surgery or Brain Damage

We tend to think of brain activity as "turning on" behavior, but equally important is the role of brain activity in "braking" behavior. In the course of brain surgery performed to treat various conditions, depending on the brain regions cut or removed, components of sexual behavior may be inhibited (by damage to a normally stimulatory region) or exacerbated, or "disinhibited" (by damage to a normally inhibitory region). Here we describe several types of brain surgery that have been performed and their effects on sexual behavior.

Frontal Lobotomy

The most extensive and controversial series of human brain-lesioning procedures (known as "psychosurgery") was carried out by the team of Walter Freeman and James W. Watts. They performed more than 550 "frontal lobotomies" in the 1930s and 1940s, mainly for the treatment of schizophrenia and psychosis, and were eventually "personally involved with 3,500 lobotomized patients" (Freeman, 1971). They "endeavored to find the best plane of section that would give the patient relief from his emotional turmoil while at the same time leaving him with enough frontal connections to restore him to useful activity." The best plane was deemed to be the transorbital (through the eye socket!). This allowed the patient to leave the hospital within two days, with no wounds to dress and a minimum of complications, both surgical and social. Freeman described the rationale for the procedure rather sardonically: "Performed during the stage of coma after electroconvulsive shock, it proved to be the ideal operation for use in crowded state mental hospitals with a shortage of everything except patients." In this 1973 publication, nothing at all was stated about the sexuality of the patients after or before the psychosurgery, but other Freeman and Watts publications did describe some effects, though with sparse detail.

In an extensive review of the function of the frontal lobes, published in 1953, Frank Ervin characterized the personality changes following frontal lobotomy in patients with psychosis: "slight impairment in ability to generalize, a tendency to perseverate, euphoria, apathy, procrastination, facetiousness, temper outbursts and distractability, impaired judgement, lack of planning for the future, and loss of creative imagination." However, he made no reference to changes in sexuality after frontal lobotomy.

In a "final report of 500 Freeman and Watts [frontal lobotomy] patients who were followed for 10 to 20 years," Freeman (1958) characterized the failures as "personality alteration . . . Lack of the sense of responsibility, boisterous laughter, rudeness and unproductive restlessness." He characterized the personality of the "successes" after the surgery as perhaps being "a bit on the 'routine' side but endowed with enough energy and imagination to make an adequate adjustment in their social lives." However, in this report he made no comment whatsoever about the sexuality of any of the 500 patients.

In their book published in 1950, Freeman and Watts characterized the effects of their frontal lobotomy surgery as follows: "Partial separation of the frontal lobes from the rest of the brain results in reduction of disagreeable self-consciousness, abolition of obsessive thinking, and satisfaction with performance, even though the performance is inferior in quality." They quoted Harvey Cushing's comment on a patient in whom he had separated the right frontal lobe from the rest of the brain to remove a tumor: "The 'defrontalized' dement, deprived of one of the principal centers of psychic life, recovers his equilibrium at the price of intellectual impoverishment, but it is better for him to have a simplified intellect, capable of elementary acts, than an intellect where reigns the disorder of subtle syntheses. Society can accommodate itself to the most humble laborer, but it justifiably distrusts the mad thinker" (Freeman & Watts, 1950).

McKenzie and Proctor (1946) noted increased sexuality in some 25 percent of their patients who had undergone frontal lobotomy. Freeman and Watts (1950) commented that "sexual behavior in the majority of cases does not seem to undergo any great alteration following prefrontal lobotomy." They did, however, describe some of their postsurgery patients in whom sexuality was affected:

> [Case 29: After the frontal lobotomy surgery] he would slap the nurse on the rump every time she got within reach . . . When he was chided for this he did not seem to be upset but rather to understand that it was something that we all disapproved of. Nevertheless, he said, "Why shouldn't I do it? I like to." On another occasion, when again chided about this act that was giving him a bad reputation among the nurses, he did not seem particularly abashed about it and more or less refused to accept responsibility.

> [Case 42: Before the frontal lobotomy surgery] when her husband came into the room or if she heard his voice—sometimes when she merely thought of him—there would be a surge of emotion . . . After operation the patient reported that her husband could come and go without causing her heart to flutter . . . and that her affection for him was just as deep, only possibly somewhat calmer.

> [Case 61:] A young woman of vivacious temperament had experienced no sexual pleasure with her husband during eleven years of married life . . . On the same evening after operation she began kissing her husband passionately.

> In summing up our observations on the relation of prefrontal lobotomy to sexual activity, it is our impression that in this respect, as well as in many others, there is an alteration in the many conflicting tendencies that were present before operation. It would seem that the postoperative inertia manifested by some patients reduces the tendency of the individual to seek sexual gratification. On the other hand, the suppression of the restraining forces may lead to a freer expression of the personality along sexual lines.

Freeman and Watts (1950) offered some bold advice to the wives of men who might get too frisky as a result of their lobotomy procedure (we can't quite tell whether, in places, they were serious or were displaying 1950s' medical humor):

> Increase in sexual appetite and performance is a common phenomenon after prefrontal lobotomy and may continue for several months. It is apt to be at a somewhat immature level in that the patient seeks sexual gratification without particularly thinking out a plan of procedure. The freedom from conflict in some cases makes for renewed delight. Sometimes the wife has to put up with exaggerated attention on the part of her husband, even at inconvenient times and under circum-

> stances when she may be embarrassed, and sometimes it develops into a ticklish situation. Patients may complain that their wives are no longer affectionate. This is easily understandable. A woman who has been the target of her husband's criticism all day long may well be excused for avoiding his embraces when bedtime comes, particularly since the sexual act is carried out only for his own gratification. Refusal, however, has led to one savage beating that we know of (Case 43), and to several separations. Physical self-defense is probably the best tactic for the woman. Her husband may have regressed to the cave-man level, and she owes it to him to be responsive at the cave-woman level. It may not be agreeable at first, but she will soon find it exhilarating if unconventional.

Overall, rather pleased with their results, Freeman and Watts (1950) conclude: "We are quite happy about these folks, and although the families may have their trials and tribulations because of indolence and lack of cooperation, nevertheless when it comes right down to the question of such domestic invalidism as against the type of raving maniac that was operated on, the results could hardly be called anything but good."

Limbic System Surgery

There are reports of "hypersexuality" produced by brain damage. The basis for the hypersexuality may be a combination of loss of discrimination of appropriate sexual objects, loss of social inhibitions, and increased sexual desire.

The classic example of hypersexuality as a syndrome was described by Kluver and Bucy and later named (by others) the Kluver-Bucy syndrome. They observed the syndrome in rhesus monkeys that had undergone surgical removal of the temporal lobes, including the amygdala, on both sides of the forebrain (Bucy & Kluver, 1955). They characterized the effects of this surgery as follows: "(1) forms of behavior indicative of an agnosia [inability to recognize or comprehend] in various sense fields [they also referred to this as *Seelenblindheit* ('psychic blindness'—e.g., a lesioned monkey handling a snake, of which it is normally fearful)], (2) strong oral tendencies, (3) an excessive tendency to attend and react to every visual stimulus, (4) profound changes in emotional behavior, (5) striking changes in sexual behavior, particularly in the form of hypersexed behavior, and (6) changes in dietary habits (e.g., eating of meat)." The "hypersexed" behavior included attempts at mating with animals of other species and inanimate objects and frequent masturbation.

Other investigators later observed similar effects of experimental lesions in or removal of the same brain region. The temporal lobe includes several different identifiable brain regions—including the pyriform cortex, the amygdala, and the hippocampus. Since these brain divisions are contiguous with each other, it is difficult to lesion one without encroaching on others. Consequently, there are some inconsistencies among the studies, especially among the different species used—rabbits, cats, and humans.

Schreiner and Kling (1953) observed aberrant sexual behavior in cats after lesioning of the amygdala and parts of the temporal lobe and hippocampus; male cats attempted to copulate with inappropriate species (hen, dog, and monkey). In other experiments on cats, J. D. Green, Clemente, and DeGroot (1957) observed what they termed "abnormal sexuality" rather than "hypersexuality" in males with lesions to the pyriform cortex, which overlies the amygdala. Some of these cats had no detectable injury to the amygdala. The lesioned male cats attempted copulation with "rats, guinea pigs, rabbits, teddy bears, anesthetized animals and even humans." The authors decided to euthanize one cat that showed repeated generalized seizures. "Accordingly, while mounted on a teddy bear, he was given a lethal dose of Nembutal [a barbiturate]. He continued to attempt coitus until he fell off the teddy bear, apparently anesthetized. When an attempt was made to remove the teddy bear, he aroused sufficiently to remount and continue for about 30 seconds, after which he died. Several animals attempted coitus with the teddy bear for periods of 20 to 45 minutes, only ceasing from exhaustion."

Beyer, Yaschine, and Mena (1964) lesioned the brains of female rabbits in the cortical region similar to that lesioned in cats in the J. D. Green, Clemente, and DeGroot (1957) study. The lesioned female rabbits showed male-type mounting behavior when placed with male rabbits, baby rabbits, cats, and guinea pigs. The rabbits that had been operated on "persisted in mounting even when viciously rebuffed by the other animals." This behavior disappeared after removal of the ovaries and was restored by estrogen injections. While male-like mounting behavior was also observed in 7 of 27 intact female rabbits, it lacked the intensity, persistence, and compulsory character found in the lesioned rabbits, and intact rabbits did not mount cats or guinea pigs.

The studies described above were inspired by a report by Terzian and Dalle Ore (1955) on a 19-year-old man with temporal lobe epilepsy that was resistant to medication. He had undergone surgical removal of the anterior portion of the temporal lobes, anterior portion of the hippocampus, and the amygdala, all bilaterally. After the surgery, he showed several manifestations resembling the Kluver-Bucy syndrome. For example, "He picked up objects . . . the same object again and again . . . ate at least as much as four normal persons . . . displayed to the doctor, with satisfaction, that he had spontaneous erections followed by masturbation and orgasm . . . became exhibitionistic . . . wanted to show his sexual organ erect to all doctors . . . [and] showed indifference [to women] in contrast with his behavior before the operation . . . Homosexual tendencies . . . were soon noticed . . . [and he practiced] self-abuse several times a day."

There are also several reports of hypersexuality in humans after surgery that encroached on the septal region of the brain. These cases involved an increase in sex-related activities rather than an increase in orgasmic activity. For example, Gorman and Cummings (1992) reported the case of a 75-year-old man living in a nursing home who developed hydrocephalus (excess fluid in the brain) and had a shunt placed in his brain to drain the fluid. After this surgery, there were many reports of the patient approaching and fondling female patients. He was noted to crawl into bed with other patients, with sexual intent, and to use sexually explicit language. (Before the shunt placement, "he had never been married, courted very little, and had never used coarse or suggestive language.") In the nursing home, he had to be constantly restrained to prevent the sexual behavior. After three years he was referred to Gorman and Cummings, who performed a CT (computerized tomography) scan and found that the tip of the shunt was lodged in the septum in the medial aspect of the floor of the lateral ventricles (the hollow inner part of the forebrain), at the junction of the frontal horns. They treated the patient with carbamazepine (an antiepileptic), haloperidol (an antipsychotic), propranolol (a beta-adrenergic blocker), and diethylstilbestrol (a synthetic estrogen) for at least one month each, with "no discernible effect on his sexual behavior." The investigators made the droll comment: "He declined shunt revision."

The same authors reported the case of another 75-year-old man who developed encephalitis followed by coma (Gorman & Cummings, 1992). He recovered, but developed ventricular enlargement, necessitating placement of a ventricular-peritoneal shunt (to drain the fluid from the brain to the abdominal cavity). Before developing encephalitis, the patient and his wife had had weekly sexual relations. After recovery from the encephalitis, "he made sexual comments toward women he encountered in the hospital, attempted to fondle the nurses, and masturbated publicly. After discovery of the hydrocephalus and shunt placement, his previously disinhibited sexual response was markedly increased and became 'disgusting' according to his wife. He became 'the man with a thousand hands,' attempting to fondle her each time she came within reach. He requested intercourse with her many times each day and also asked that she have sex with other men while he watched, an interest never previously expressed." By the time Gorman and Cummings assessed him, this excessive sexual interest had been present for two years. Again, the investigators state: "The patient refused shunt revision."

B. L. Miller et al. (1986) noted that "hypersexual behavior following brain injury is uncommon but when seen is often associated with basal frontal or diencephalic lesions [regions of the forebrain] . . . patients manifested disinhibited public expression of their increased drive. This sexual disinhibition was often associated with a general disinhibition of behaviour." They reported the case of a 39-year-old man in the hospital who publicly masturbated and attempted to have intercourse with his wife and with female nurses in front of his three roommates. Hemorrhage was found in his basal frontal areas bilaterally, with involvement of the septal region. The authors also described cases of lesions in other brain regions related to socially inappropriate sexual behavior.

A 59-year-old man underwent surgical removal of a subfrontal meningioma (a tumor in the meninges, the membranes that encapsulate the brain) revealed by a CT scan (B. L. Miller et al., 1986). The authors noted that "after surgery his desire for sexual activity increased from once per week to 1–4 times per day. Intercourse frequently lasted longer than 1 hour and he had some difficulty achieving orgasm . . . Two years after surgery he became increasingly preoccupied with sex and developed a manic syndrome. He was admitted to hospital where he publicly masturbated and sexually propositioned both male and female patients and staff (he had a past history of previous homosexual contacts)."

The same authors described the case of a 34-year-old man with a glioma (tumor of the glia, which are nonneuronal brain cells) involving the thalamus, hypothalamus, ventral midbrain, and pons (B. L. Miller et al., 1986). The man began to make sexual proposals to his 7-year-old daughter and her friends, made increasingly more public sexual advances toward young children, and frequently embarrassed his wife by showing pornographic pictures to visitors at their home. He was arrested for propositioning children in his neighborhood.

In the 1970s in Germany, attempts were made to control pedophilia and "sexual delinquency" through the use of unilateral surgical removal of the ventromedial or medial anterior hypothalamus (Dieckmann et al., 1988). The authors described the results of one such surgical method: "Following a stereotactic ventromedial hypothalamotomy not only are the dynamic aspects of sexuality, such as compulsion and impulsivity, diminished, but also organic components, which have to do with completion of the sex act. In contrast, the structure of the patient's sexual organization remains unchanged: A pedophilic character, for example, is still retained. However, it is possible for the subject to adapt his sexual behavior to the specific conceptions and expectations of our society."

Based on all of these cases, we can conclude that the "higher functions" of the brain, such as the learning of socially acceptable behavior, are a complex process involving finely tuned inhibitions. When the frontal regions are disconnected from the rest of the brain, that complexity is lost and, with it, the fine tuning of social graces. Prominently disinhibited is sexual behavior, along with other complex sociocultural behavior patterns. Perhaps there are parallels in normal development. These finely tuned sociocultural patterns are the last to emerge in the developing child, and the first to be lost in senile dementia. Thus, they are the most complex, and the most vulnerable, brain functions.

22

Imaging the Brain during Sexual Arousal and Orgasm

Starting about two decades ago, two technological breakthroughs, best known by their acronyms, PET and fMRI, made it possible to generate pictures of the activity in all parts of our brain simultaneously, while we are awake—something never before possible. This technology has rapidly surpassed previous methodologies and revolutionized our ability to understand how our brains function in cognition and behavior.

What fMRI and PET Tell Us

Neuroscientists have used a variety of technologies to provide insights into the neural basis of orgasm. Each method has its unique advantages and limitations. Functional brain imaging such as PET (positron emission tomography) and fMRI (functional magnetic resonance imaging) provided the first evidence of the brain areas involved in naturally induced orgasm in humans. The advantage of both of these methods is that they provide a three-dimensional map in awake humans of brain regions that are active during a condition of interest (e.g., vaginocervical self-stimulated orgasm) relative to a control condition (e.g., unstimulated resting, or vaginocervical self-stimulation before or after orgasm). The PET and fMRI methods have an advantage over EEG (electroencephalography) in that EEG does not provide information on localized activation deep in the brain. Both PET and fMRI also have an advantage over the use of implanted electrodes that record multineuronal action potential (impulse transmission) activity in that neither fMRI nor PET is invasive into the brain.

The major limitation of the fMRI and PET methods is that rather than providing a measure of neural activity per se, they measure a hemodynamic (blood-related) response that is an indirect measure of neural activity. Typically, in PET, the subject is injected intravenously with water labeled with a radioactive isotope of oxygen, oxygen-15. Because oxygen-15 has a half-life of only 2 minutes, it must be prepared in an on-site cyclotron and incorporated into water in a laboratory next to the room where the subject awaits the injection and PET. The distribution of radioactivity in the brain is then mapped in three dimensions by the PET scan. Blood flow normally increases locally in regions of increased neuronal activity, so there is a relative increase in the amount of radioactivity (i.e., oxygen-15) per unit time in those regions. It is this relative change that provides the data in this type of PET. Because of the 2-minute half-life, subsequent injections of labeled water are typically administered at 15-minute intervals (e.g., intervals corresponding to alternating control and experimental stimulus conditions).

The limitations of the PET method include the need for invasive intravenous injections of radioactivity and the complexity and temporal constraints of coordinating the on-site production of oxygen-15, immediately using it in the synthesis of water, immediately injecting the water intravenously, waiting for the water to be distributed to the brain, and timing that pulse of radioactivity to coincide with the orgasm—a rather heroic effort on the part of, at least, the experimental subject (e.g., Komisaruk, Whipple, Gerdes, et al., 1997; Komisaruk, Whipple, Crawford, et al., 2002; Whipple & Komisaruk, 2002). Furthermore, since the region of increased radioactivity is relatively diffuse, the method is better suited to experiments in which a relatively large brain region is expected to be involved, such as the neocortex or basal ganglia, rather than much smaller regions such as specific groups of nuclei in the brainstem.

The fMRI method is also based on the increased blood flow that supplies neurons when they become active. When the neurons are more active, they use more oxygen. Two processes result: first, oxygen is removed from the hemoglobin in the blood that supplies the neurons, and second, there is a compensatory increase in blood flow to the region, carrying more oxygenated hemoglobin (oxyhemoglobin) to the neurons (Ogawa et al., 1990). It is likely that release of local factors by neurons when they become active relaxes the blood vessels supplying those neurons, thus increasing the local blood supply. One possible mediator could be nitric oxide, which is released by neurons when they are active and relaxes the smooth muscle of blood vessels. It is unlikely that dilation of the blood vessels (vasodilation) is mediated by the innervation of those vessels, because this would make the system far more complicated. In other words, local vasodilation by substances released by neurons when they are active seems to be the simplest explanation for the hemodynamic changes measured by fMRI (and PET). Several studies have shown that while various factors can influence the fMRI signal, the magnitude of neuronal activity is correlated with the magnitude of the fMRI signal (for a review, see Arthurs & Boniface, 2002).

The fMRI method does not use radioactive material, but depends on the difference in magnetization between oxyhemoglobin and deoxyhemoglobin (hemoglobin that has given up its oxygen). The more active the neurons, the more oxygen they withdraw from the blood and the more oxygenated blood is resupplied to that neuronal region. Because the magnetic property of the iron in the hemoglobin is affected by whether it is combined with or depleted of oxygen, these two processes—oxygenation and deoxygenation—create a perturbation in the local magnetic environment. This is mapped in three dimensions, providing the data for the fMRI (Ogawa et al., 1990). The resolution of fMRI is sharper than that of PET, to the extent that fMRI can map the location of specific motor and sensory clusters of neurons (i.e., specific cranial nerve nuclei) in the brainstem that are activated by specific motor or sensory tasks (Komisaruk, Mosier, et al., 2002).

There are various strategies for analyzing fMRI or PET data, such as comparing the observed activity in a region to that of other regions concurrently in the same brain "slice" (the plane of the brain being viewed) or comparing it to activity when sensory stimulation is not being applied. Another consideration is the type of analytical strategy to use. One strategy is to start without a prior hypothesis and analyze where activity in one brain region differs significantly from activity in another. An alternative strategy is to make an a priori selection of specific "regions of interest" to be analyzed and then compare the activity under stimulation and nonstimulation conditions. A third consideration is where to set the threshold at which activation is considered significantly greater than in the selected controls. If the threshold is too stringent, areas of activation will be missed; if it is too low, activity in the "background" will obscure activity in the specific regions of interest (e.g., Poline et al., 1997).

Imaging Brain Activity during Arousal and Orgasm in Women

Komisaruk, Whipple, and colleagues initially used the fMRI method to test the hypothesis that the vagus nerves convey vaginocervical

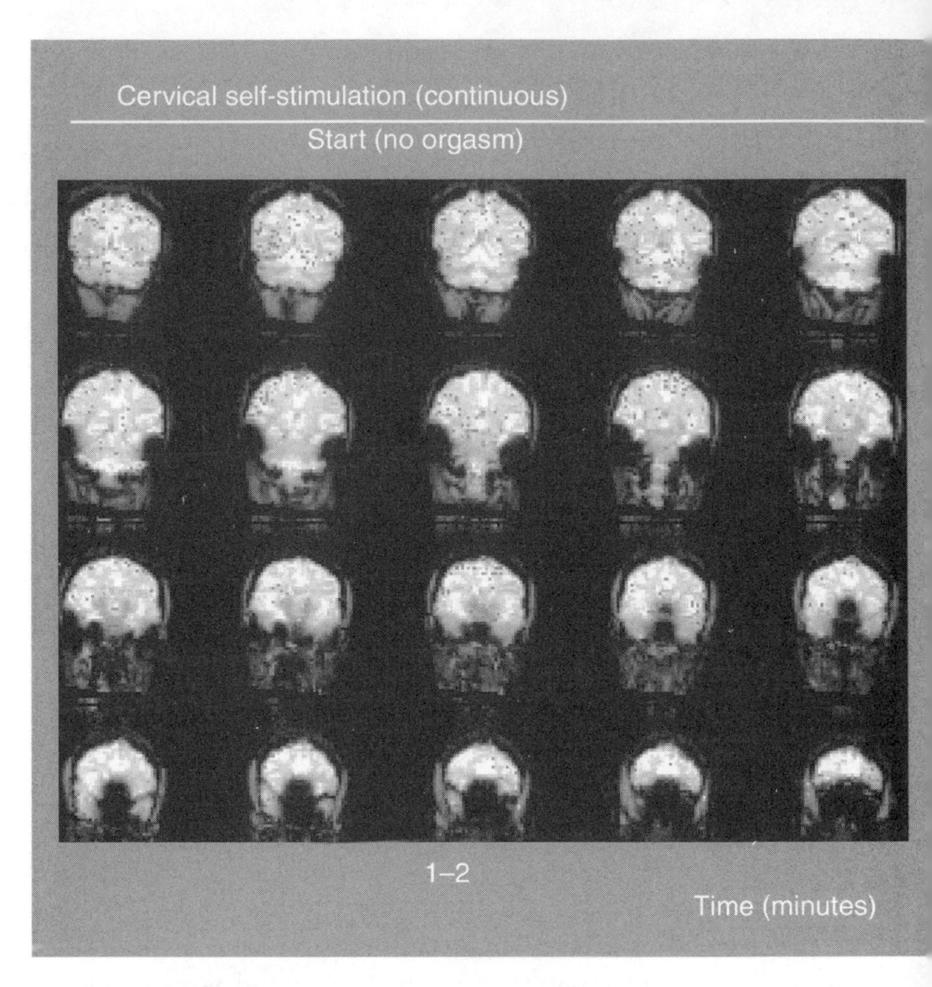

A series of fMRI images showing successive "slices" through the brain from front (lower right side of each panel) to back (upper left in each panel) of a woman during vaginocervical self-stimulation that was continuous for four minutes, before (left panel) and during (right panel) an orgasm. Regions of brain activation appear as darkened pixels

sensation directly to the brain, by ascertaining whether the nucleus of the solitary tract (NTS) is activated by vaginocervical stimulation in women with complete interruption of the spinal cord. The NTS is the brain region in the medulla of the lower brainstem to which the sensory component of the vagus nerve projects. The fMRI studies showed that the NTS is activated by this stimulation, thus supporting the hypothesis. While performing the vaginocervical self-stimulation, some of the women experienced orgasms,

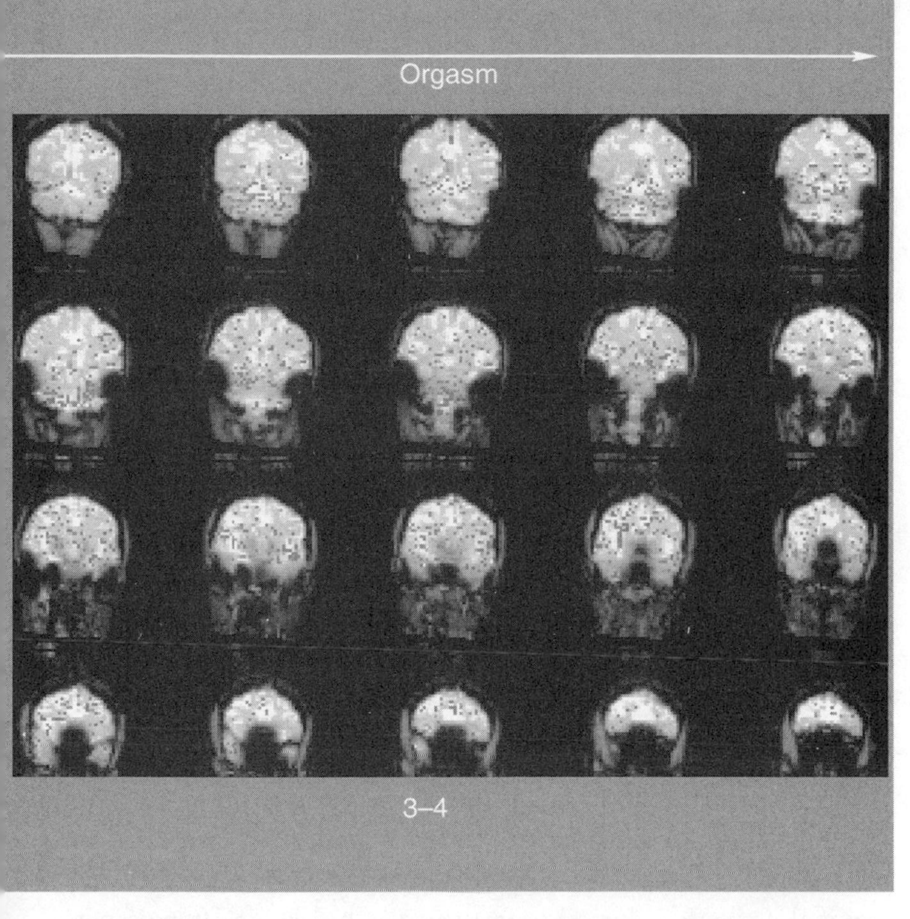

(dots). Note the much greater number of activated pixels, hence activation, throughout the brain during orgasm than before orgasm. See the text for further details. (Adapted from Komisaruk, Whipple, Crawford, et al., 2004, with permission from *Brain Research*)

and fMRI recordings were made before, during, and after orgasm (Komisaruk, Whipple, Crawford, et al., 2002; Whipple & Komisaruk, 2002; Komisaruk, Whipple, Crawford, et al., 2004; Komisaruk & Whipple, 2005).

The brain regions activated during orgasms induced by vaginocervical self-stimulation include the hypothalamus, the limbic system (including the amygdala, hippocampus, cingulate cortex, insular cortex, and the region of the accumbens-bed nucleus of the

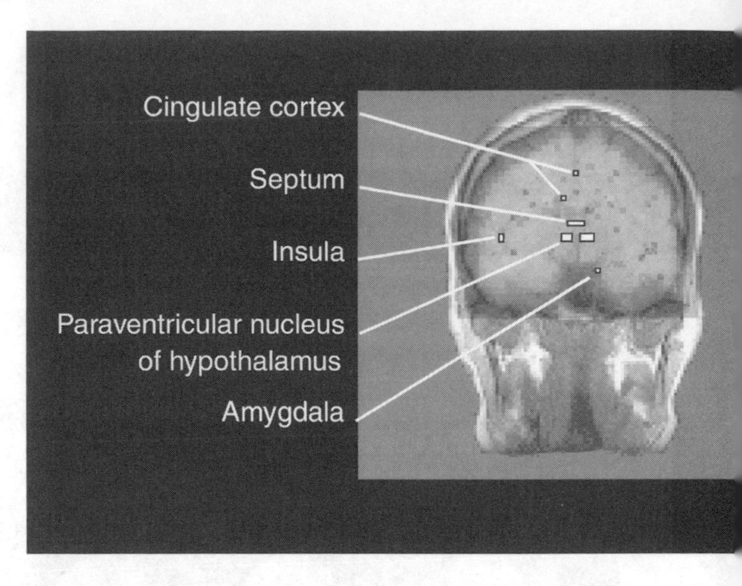

Two fMRI images of different regions of the forebrain of a woman during an orgasm produced by vaginocervical self-stimulation. Some of the activated pixels have been highlighted and are labeled. See the text for discussion of the cognitive roles of these brain regions and

stria terminalis–preoptic area), the neocortex (including parietal and frontal cortices), the basal ganglia (especially putamen), and the cerebellum, in addition to the lower brainstem (central gray, mesencephalic reticular formation, and NTS). Differences between regional activation during orgasm versus before or after orgasm suggest that areas related to vaginocervical stimulation–induced orgasm include the paraventricular area of the hypothalamus, the amygdala, the anterior cingulate region of the limbic cortex, and the region of the nucleus accumbens (Komisaruk, Whipple, Crawford, et al., 2004).

While there is no evidence of orgasm in female rats, some of the same-named brain regions become activated during mating or vaginocervical stimulation. Thus, using the c-fos immunocytochemical method (which makes visible a specific protein produced in a neuron when stimulated) in rats, activation was reported in the amygdala, paraventricular nucleus of the hypothalamus (PVN), medial preoptic area, midbrain central gray, and—based on the

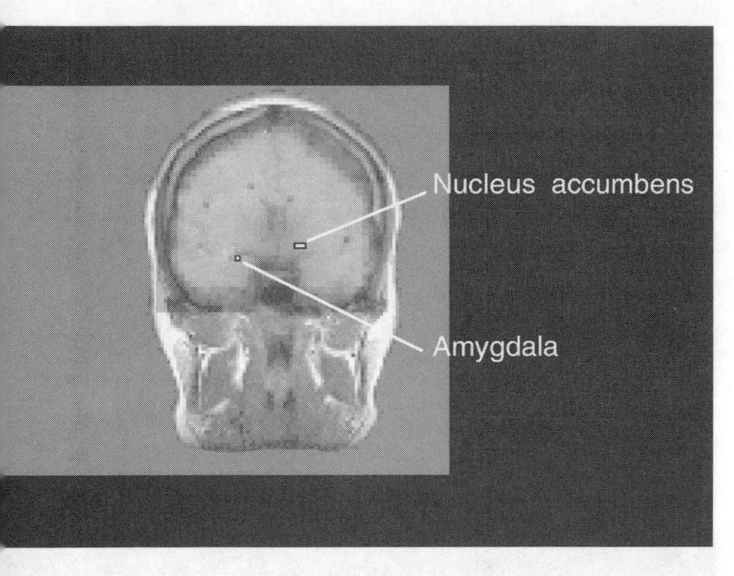

their contribution to orgasm. (Adapted from Komisaruk, Whipple, Crawford, et al., 2004, with permission from *Brain Research*)

local release of dopamine—the nucleus accumbens (*amygdala:* Rowe & Erskine, 1993; Tetel, Getzinger & Blaustein, 1993; Wersinger, Baum & Erskine, 1993; Erskine & Hanrahan, 1997; Pfaus & Heeb, 1997; Veening & Coolen, 1998; *PVN:* Pfaus & Gorzalka, 1987; Rowe & Erskine, 1993; *medial preoptic area:* Tetel, Getzinger & Blaustein, 1993; Wersinger, Baum & Erskine, 1993; Erskine & Hanrahan, 1997; *midbrain central gray:* Tetel, Getzinger & Blaustein, 1993; Pfaus & Heeb, 1997; *nucleus accumbens:* Pfaus et al. 1995).

The finding of activation in the region of the PVN (Komisaruk, Whipple, Crawford, et al., 2004) is consistent with reports of oxytocin release during orgasm. That is, the PVN neurons secrete oxytocin, which is stored in the posterior pituitary gland (Cross & Wakerley, 1977). Vaginal or cervical stimulation releases the oxytocin from the posterior pituitary into the general bloodstream—a response known as the "Ferguson reflex" (J. K. W. Ferguson, 1941)—and orgasm releases the oxytocin into the general blood-

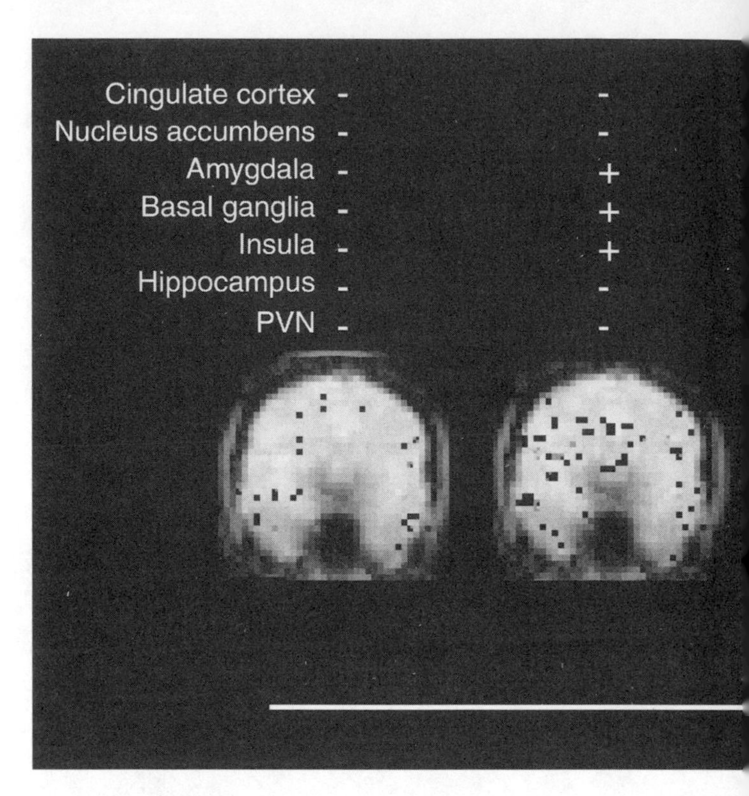

Sequential fMRI images of one "slice" of the forebrain of a woman during the development of orgasms in response to continuous vaginocervical self-stimulation. Each "slice" shows the activity in the brain during a one-minute period. The woman experienced the orgasms during the last two one-minute periods. Seven forebrain

stream (Carmichael, Humbert, et al., 1987; Carmichael, Warburton, et al., 1994; Blaicher et al., 1999).

A salient and reliable feature of the brain regions activated during orgasm is activation of the cerebellum. The cerebellum regulates muscle tension via the "gamma efferent system" and it receives proprioceptive information (Netter, 1986). (Proprioceptors are sensory receptors in muscles and joints.) Mould (1980) emphasized the role played by the gamma efferent system in generating the muscle tension characteristic of orgasm. The cerebellum provides the main cerebral control of this system. "Orgasm as a total body response," Mould stated, ". . . almost certainly necessitates

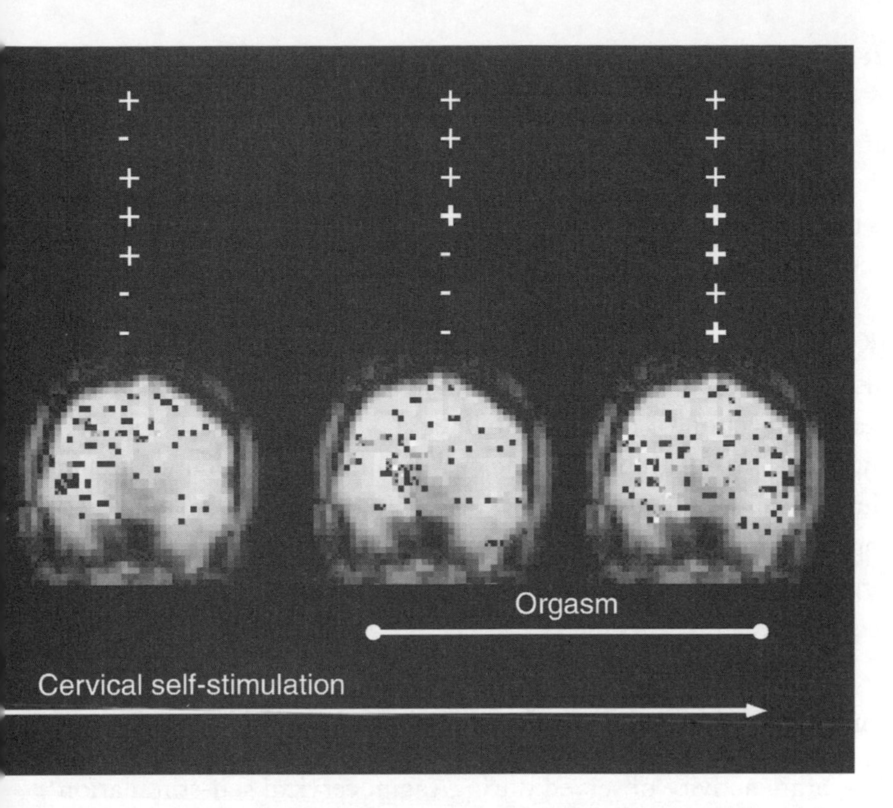

regions are listed at the left (PVN, paraventricular nucleus of the hypothalamus); a minus sign indicates that the region was inactive during that minute; a plus sign, that the region was activated during that minute. A bold plus sign indicates that several pixels in the region were activated. (Adapted from Komisaruk, Whipple, Crawford, et al., 2004, with permission from *Brain Research*)

cerebellar . . . organization and coordination." Muscle tension can reach peak levels during orgasm (Masters & Johnson, 1966) and contribute to sensory pleasure (Komisaruk & Whipple, 1998, 2000). While it is likely that the cerebellum plays a significant motoric (muscle movement) role in orgasm via the gamma efferent system, it is tempting to speculate that it also plays a significant perceptual/cognitive-hedonic role.

It is of interest that the insular and anterior cingulate cortices are active during orgasm, because both of these areas have been reported to be activated during response to pain (Casey, Morrow, et al., 2001; Bornhovd et al., 2002; Ploner et al., 2002). This suggests an interesting local interaction between the regions of the brain that are activated during both pain and pleasure. Further research is needed to compare, within the same individual, the brain regions activated during pleasure with those activated during pain.

The region of the nucleus accumbens is activated during orgasm (Komisaruk, Whipple, Crawford, et al., 2004). This brain region also shows fMRI activation during the "rush" induced by an intravenous injection of nicotine (Stein et al., 1998). And as measured by fMRI, activation of the nucleus accumbens in humans occurs during feelings of euphoria produced by a cocaine "rush" and feelings of craving (Breiter et al., 1997). The fMRI findings during orgasm suggest a role for the nucleus accumbens in mediating orgasmic pleasure in women.

Imaging Non–Sensory-Induced Orgasms

The brain activity observed during vaginocervical self-stimulation necessarily includes the brain activity that is generating the arm and hand movements producing the self-stimulation as well as the brain activity that is the sensory response to the stimulation. Rather than trying to discount such motor and sensory representation from the orgasm records, and to obviate making assumptions about which brain regions to discount, Komisaruk and Whipple (2005) have studied women who can experience orgasm by thought alone—without any physical self-stimulation or other form of stimulation. In an earlier study, they compared the autonomic responses of ten women who were able to experience orgasm by thought (Whipple, Ogden & Komisaruk, 1992) under two conditions in each woman: one during genital self-stimulation–induced orgasm and the other during thought-induced orgasm. The magnitude of each parameter measured—heart rate, blood pressure, pupil dilation, and pain threshold—approximately doubled during orgasm compared with the initial resting baseline. The significant increases in these physiological measures occurred during the thought condition as well as the physical genital self-stimulation condition. The women described the imagery they experienced during the thought-induced orgasms variously—in some cases erotic, others pastoral, and still others abstract, such as "energy flow" repeatedly ascending and descending the body axis.

Preliminary findings from fMRI studies indicated that in thought-induced orgasms, as in orgasms induced by vaginocervical self-stimulation, the regions of the nucleus accumbens, PVN, hippocampus, and anterior cingulate cortex are activated. This suggests that activation of these brain regions is rather "specifically" related to orgasm, in the sense that it is not related to the brain control of the efferent and the consequent "reafferent" activity that generates orgasm via hand movement in response to genital self-stimulation. (Initial stimulation ["afference"] produces intensified contraction of smooth and/or voluntary muscles, which generates increased sensory input from these muscles ["reafference"], which intensifies the pleasurable feeling of orgasm.) The amygdala was not activated during the thought orgasms, suggesting that it is perhaps closer to having a *genital sensory* role in orgasm, while the other regions activated may be more *cognitively* related to orgasm.

Other Brain Imaging Studies in Women and Men

Several other studies have imaged the brain during sexual arousal and orgasm in women and men. It is difficult to derive generalizations from these studies because their procedures differ substantially, there is no standardization of criteria for the threshold of what constitutes a significant activation, and the different recording methods have greater or lesser sensitivity and regional resolution of brain regions. Nevertheless, despite these caveats, we can draw several interesting conclusions from these findings, which could not have been obtained with earlier methodologies.

Several studies show that the dopaminergic "reward" system seems to be activated during sexual arousal and orgasm. This is

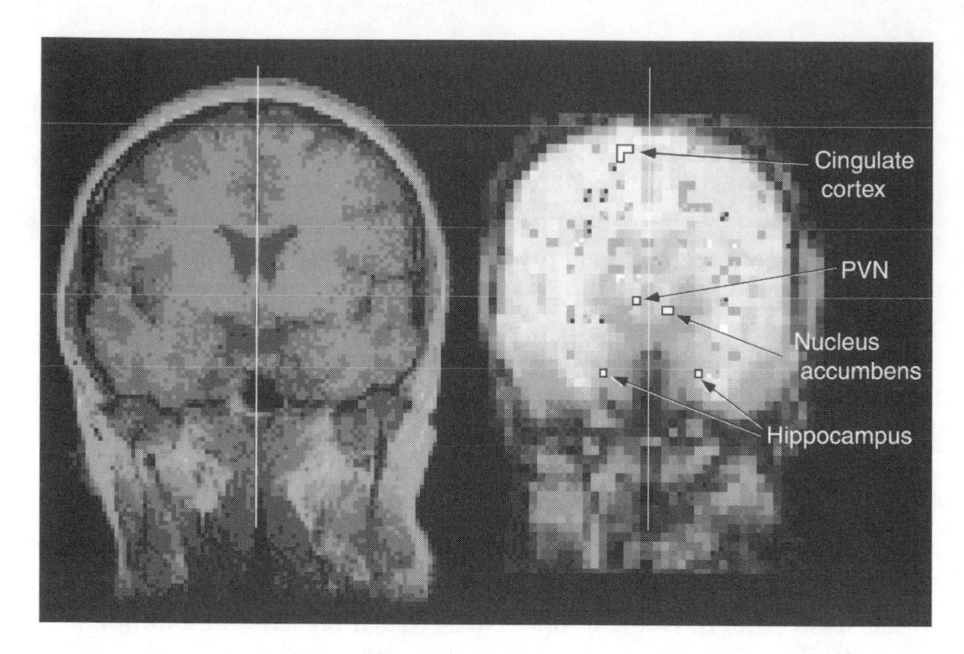

An fMRI image (*right*) of the forebrain of a woman during an orgasm elicited by thought only, without physical stimulation. The anatomical image of the brain region where the fMRI was recorded is also shown (*left*). Note the activation of some of the same brain regions as during orgasm elicited by vaginocervical self-stimulation. Since there was no arm or hand movement, or genital sensory stimulation, the brain regions activated here are more closely related to orgasm per se, without being affected by the other factors. PVN, paraventricular nucleus of the hypothalamus.

supported by the fMRI studies showing that the nucleus accumbens region is activated during orgasm in women (Komisaruk, Whipple, Crawford, et al., 2004). Holstege et al. (2003), using PET, found that the mesodiencephalic area, which projects to the same (nucleus accumbens) region, was activated in men during orgasm. In fMRI studies, Aron et al. (2005) found that men and women who were "intensely in love" showed activation in the ventral midbrain and the caudate nucleus when observing pictures of their beloved. These regions are also components of the dopaminergic system.

The dopaminergic system was also implicated in a pharmacological study of brain areas activated during penile erection (Bornhovd et al., 2002). Brain activity was measured with PET in men with erectile dysfunction, who were receiving the dopaminergic stimulatory drug apomorphine, while they watched a sexually stimulating video sequence. The investigators found a significant correlation between the degree of penile rigidity (measured with the commercial RigiScan device) and increased cerebral activity in the anterior cingulate cortex and right prefrontal cortex, in addition to decreased activity in the temporal cortex.

Park et al. (2001) found that the visual ("occipital") cortex in women was much more highly activated when the women observed erotic films than when they observed nonerotic films. Thus, rather than the visual cortex being a "passive visual projection screen," its activity evidently is modulated by the cognitive-emotional content of what is being observed.

In a study by Karama et al. (2002), men were shown erotic film segments and were asked to rate their subjective level of arousal on a visual analog scale—a graded scale much like a ruler on which the subject places a mark that represents his arousal level, somewhere between zero and maximum. The higher the men's self-reported level of sexual arousal, the greater was the fMRI activity recorded in their hypothalamus. Hypothalamic activation was also reported in an fMRI study of men watching explicit erotic videos while penile tumescence was measured (Arnow et al., 2002). During sexual arousal, the cingulate gyrus, insula, basal ganglia (claustrum, caudate, and putamen), midtemporal gyri, and premotor areas all were activated. In another self-report study, by Redoute et al. (2000), men rated their level of sexual arousal as their brain activity was measured by PET. The brain area where activation showed the best correlation with the men's perceived level of sexual arousal was the anterior cingulate gyrus in the forebrain. The anterior cingulate was also activated in women during sexual arousal, in addition to the amygdala (Maravilla, 2006), consistent with the findings of Komisaruk, Whipple, Crawford, and colleagues (2004).

To our knowledge, the first report on brain imaging of men during orgasm was a PET study from Finland (Tiihonen et al., 1994).

The investigators found an increase in activity in the (right) prefrontal area. They referred to a parallel with an earlier report of "hypersexuality" in men with a lesion or disconnection of the prefrontal cortex from the rest of the brain (B. L. Miller et al., 1986). This raises a question as to whether diminished function of the prefrontal cortex during or leading up to orgasm in men may be related in some way to "wild abandon," "risky sex," or the type of disinhibited sexuality that may occur after frontal lobotomy.

Some studies have compared brain activity in men and women during sexual arousal. Two research groups reported that the hypothalamus showed greater activation in men than in women as they watched erotic visual stimuli. Karama et al. (2002) compared the fMRI of men and women watching erotic film segments. The men showed greater activity than the women in the hypothalamus and thalamus. Both men and women showed activation in the amygdala and ventral striatum and in the anterior cingulate, insular, orbitofrontal, medial prefrontal, and occipitotemporal cortices. In another study comparing brain activity in men and women as they observed erotic photographs, Wallen and colleagues (Hamann et al., 2004) reported that fMRI activity in the hypothalamus, amygdala, and hippocampus increased more in men than in women, whereas striatal regions (caudate and nucleus accumbens) were activated similarly in both men and women.

In two PET studies, the posterior hypothalamus was activated in men presented with visual sexual stimuli (Stoleru et al., 1999; Redoute et al., 2000). This region of activation was not reported in the similar studies described above. Redoute et al. (2000) drew a parallel to an earlier report that in male rats, posterior hypothalamic stimulation induced copulatory behavior and bar pressing (an activity that rats are trained to perform to receive a reward) for the opportunity to copulate (Caggiula, 1970).

Imaging Brain Activity during Orgasm—What Does It Explain?

The recent advent of high-tech brain scanners—PET and fMRI—allows us, for the first time, to recognize regions of activation in all parts of the living, awake, human brain in relation to specific stimuli and responses. As a result, a veritable land rush has developed among investigators in many fields of research, each exploring the uncharted terrain of brain activity in relation to myriad stimuli, cognitive processes, and responses.

For those of us who have availed ourselves of this great technological opportunity, after we have conducted our studies and succeeded in identifying localized regions of brain activation—such as during orgasm—we find ourselves in a quandary. How are we to understand what the specific brain regions *do,* such that this information would give us insight into their unique contributions to orgasm? The imaging is something we like to do, but what does it explain?

We might first ask what other investigators have discovered about the roles of specific brain regions that we know to be activated (or deactivated) during orgasm. For example, we know that the insular cortex is activated during orgasm, and investigators in the area of pain research have found that the insular cortex is activated during painful stimulation. So in this case, the insular cortex is a place of both pain and pleasure. How could it be (or do) both? Could this region have some property that is common to both pain and pleasure, perhaps emotional *expression*—the contorted facial expression of anguish in pain but also of impending orgasm—separate from the actual different *feelings* of pain versus pleasure?

We run the risk of creating "just so" stories—tales created to explain the origins of things—to try to make sense of disparate research findings. By reviewing some of the research findings by investigators who do not study sexual response or orgasm but who have looked at the "same-named brain structures" that are active during sexual response or orgasm, we might shed light on what is happening in the brain. An important qualifier is "same-named," describing areas of the brain that have been named by researchers. Many of these "same-named" brain regions occur in a variety of species—for example, a cerebellum is found in a human and in a rat.

Within a particular same-named brain structure, it is well-established that different regions and even different neurons in the same region perform very different functions. An important caveat, therefore, is that the mere fact that the activity of a particular brain region is noted during the occurrence of two very different functions (e.g., pleasure and pain) does not *necessarily* mean that the two functions are in some way functionally related. But this is certainly a reasonable working hypothesis, one that can be examined with further study.

It is also plausible, of course, that different neurons in a particular brain region are involved in the two different processes—pain and pleasure. If we can understand the function of the various parts of the brain that are activated during orgasm, we can, ultimately, generate a unified concept of what makes orgasm feel good. Getting back to the question of what brain imaging tells us, if we understand what each of the many different brain components contributes to orgasm, perhaps we will begin to understand how the neurons in the brain generate the feelings of pleasure during orgasm—and even, perhaps, understand the more general areas of pleasure, pain, sensory perceptions, memories, and thoughts. In other words, these studies are part of science's attempt to unravel the conscious experience.

In the next chapter we describe our basic understanding of the role, in other contexts, of the brain regions we have discussed here, as a means of trying to understand what each structure may contribute to orgasm. If that is not your cup of tea, you may want to move on to the concluding chapter, on the greatest mystery of all: how do our neurons produce consciousness?

23

The Cast of Characters

How Brain Components Contribute to Orgasm

If we want to understand how, let's say, a digital camcorder works, it is helpful if we already have some understanding of the elements of what a tape recorder does, how an image can be created by a series of dark and light dots on a television screen, and how a series of still pictures can generate the illusion of movement in a movie. While that prior understanding gives us hints as to how the camcorder works, the camcorder has an "emergent" property—something more than just those elements put together. Metaphorically, in our understanding of how the brain produces orgasm, we are at about the stage of knowing something of how the tape recorder, television, and movie camera work. The "camcorder"—orgasm—already exists, but we don't know exactly how it works. We "just"

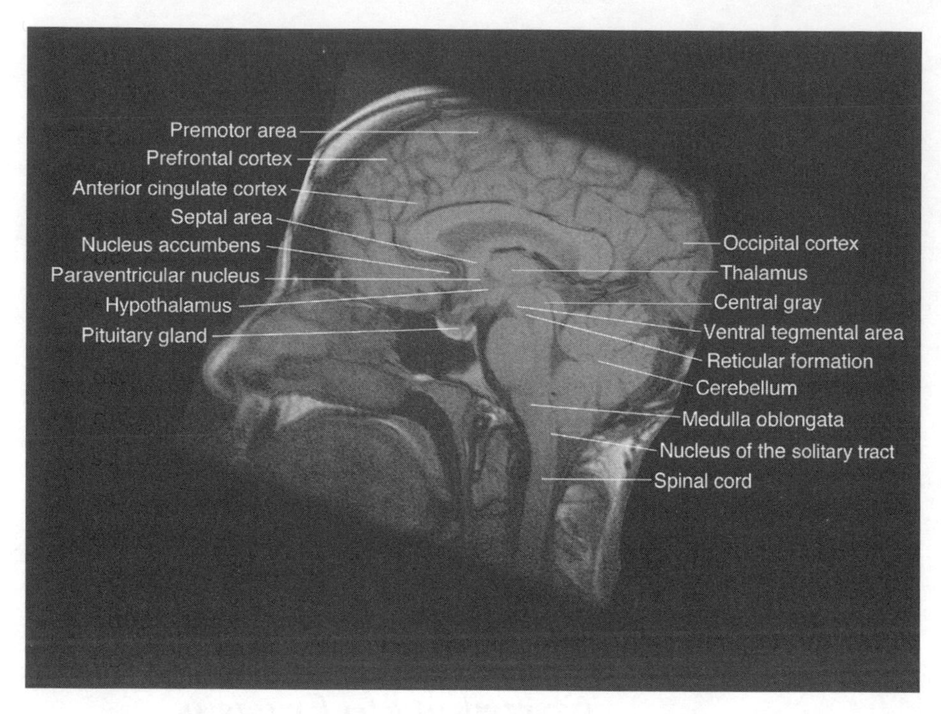

A brain viewed with magnetic resonance imaging. This midline sagittal (side) view, facing left, shows some of the brain structures mentioned in the text. Some brain structures, including the amygdala, hippocampus, and basal ganglia, lie behind the plane of the paper and are not visible at this level. (Image courtesy of Dr. Wen-Ching Liu, Department of Radiology, New Jersey Medical School, University of Medicine and Dentistry of New Jersey)

have to figure out how the roles of the brain components that we do understand are involved in orgasm, add up to an orgasm, and produce its unique properties. Thus, by reviewing the properties of some of these brain components we might be able to figure out how they contribute to orgasm. This is a basic strategy in behavioral neuroscience, and the best we can muster in our present state of knowledge.

Structure of the Limbic System

Several of the brain parts that are activated during orgasm are components of what is loosely known as the "limbic system." *Limbic* is derived from the Latin *limbus,* which means "edge, border, or fringe." It refers to the structures that surround the region where the cerebral hemispheres are attached to the brainstem. The original concept of a limbic system was formulated in 1878 by the French neurologist Paul Broca as the "Great Limbic Lobe," an anatomical description of distinct brain structures that are adjacent to each other, forming a sort of ring. This evolved into the "limbic system," with more connections. The concept has now evolved into a chimera that is part anatomically defined and part functionally defined.

Reduced to structural basics—that are evident in the embryonic human brain before it undergoes its great thickenings, bends, and folds—the brain resembles, fancifully, a Mickey Mouse balloon. The two ears represent the two cerebral hemispheres, and Mickey's head represents the brainstem, which incorporates the thalamus, hypothalamus, and lower brainstem, including the cerebellum. The entire brain is hollow, having started out embryonically as a tube; it is as if the walls of the balloon undergo great thickening, while retaining the hollow form into adulthood. In the brain, the hollow part is termed the "ventricles." Another way of thinking of the brain structure is as two hollow short-stem button mushrooms on a hollow log. The two cerebral hemispheres are the two hollow mushrooms, attached to the sides of the hollow log at its front end, with the hollow parts continuous throughout the three sections.

In this fanciful model, the "limbic lobe" would be, in large part, the circular underside of the mushroom caps. In the brain, looking at the circular underside of the mushroom cap as a clock face, the amygdala is at about 7 o'clock, the hippocampus at about 3 to 6 o'clock, and the cingulate cortex at about 9 o'clock, up and over the top of the clock, to 2 o'clock. These metaphors are not so outlandish when considering the names of some brain parts: *hippocampus,* Latin for "seahorse," which the brain structure resembles; *amygdala,* Latin for "almond," which this structure resembles; or *inferior olive, inferior* referring to location ("under"), not a value judgment. One of the early extensions of the notion of a limbic lobe was the concept that it functions as a "rhinencephalon" or "nose brain" (i.e., "brain for the sense of smell"), on the basis that the olfactory sensory pathways project to the cortical region overlying the amygdala.

Papez in 1937 grouped some of the structures of the limbic lobe into a different "system," incorporating the hypothalamus and thalamus, serially connected in a ring formation. This system projects from the hippocampus—more recently corrected to "hippocampal formation," because it is the "subicular" complex just lateral to the hippocampus proper that projects—to the mammillary bodies of the posterior hypothalamus, from there to the anterior nucleus of the thalamus, then to the cingulate cortex, to the entorhinal cortex, to the hippocampus, and back to the mammillary bodies. Later investigators termed this the "Papez circuit." Lesions or stimulation of various components of this system were found to have a variety of effects on "emotional" behavior, such as pathological crying, laughing, or aggression.

Incorporation of the hypothalamus into this chimera of "limbic lobe," "rhinencephalon," and "Papez circuit" created a conceptual *gemisch* that led Paul MacLean in 1952 to propose an extension of the concept of limbic *lobe* to limbic *system*. He claimed that the "limbic" constituents varied, depending on which expert listed them, but all now agree that the hippocampus, amygdala, and cingulate gyrus are included in the limbic system. MacLean also included areas to which these constituents connect: the septum, hypothalamus, anterior thalamic nuclei, and components of the basal ganglia. He proposed the generalization that the limbic lobe itself—in humans, more similar to the brain structures of nonhuman vertebrate species than to the human cortex, and hence considered more "primitive"—subserves emotional ("feelings") functions, whereas the more lateral neocortex, highly developed in humans, subserves more cognitive ("intellectual") functions (MacLean, 1952, 1955).

According to Nauta and Feirtag (1986), "the modern criterion for membership in the limbic system is a synaptic proximity to the hypothalamus." The limbic system plays a role in motivated behavior, as suggested by John French's notorious "4-F's—Feeding, Fighting, Fleeing, and Mating" lecture. These behavior patterns are profoundly influenced by the hypothalamus. The hypothalamus, in addition to regulating these functions (as well as hormone secretion), has been termed "the head ganglion of the autonomic system." It controls the physical plant of the body—heart rate, blood pressure, temperature, sweating, salivation, tear secretion, secretions and movements of the gut, vomiting, urination, defecation, and, of course, uterine contractions, penile erection, and ejaculation.

Connection of the limbic lobe to the hypothalamus involves the hippocampal formation output pathway—the fornix. To mix our metaphors, connecting the bottom of the mushroom button to the log, the fornix (Latin for "arch") pathway and hippocampus from which it emerges take the form of a pair of curved ram's horns, with the hippocampal tip of the horns almost impaling the amygdalae.

If you are having difficulty visualizing these structures, you are in good company. This part of the brain is particularly difficult to visualize because of its complex three-dimensional, spiral arrangement. We have provided this detail so as to introduce (among others) the brain structures that are activated during sexual arousal, orgasm, and other "motivational" states.

Motivation refers to the neural mechanisms that modify the response to a constant stimulus. In other words, hunger and satiety are considered motivational because they modify the eating response to food that is provided continuously. By contrast, the eyeblink reflex is not motivational because it is a relatively immutable response to an irritating stimulus. *Relatively* is an important caveat, because modulation of even the "simplest" reflexes—such as the knee-jerk reflex, which involves, at minimum, only one synapse between sensory neuron and motor neuron—can be modulated by voluntary action (imagine there's an antique vase just in front of your leg as its knee-jerk reflex is being tested).

We should use the term *limbic system* with another important caveat. While the term is still popularly used—and we use it in this book—the eminent neuroanatomist Larry W. Swanson (1999) notes its limited usefulness:

> The term limbic system is now used in so many ways by different investigators that it is no longer useful. When it is used, it most commonly refers to that part of the forebrain involved in emotional expression, or more broadly, in visceromotor mechanisms. However, it is now clear that while this conforms to the original concept of MacLean, the components of the limbic system are involved in cognitive and somatomotor control systems as well, and there are no clear, or even general, anatomical boundaries to this system as originally defined . . . the term limbic system has clearly outlived its usefulness, but nothing has emerged to take its place.

Swanson suggests a dichotomous division into *medial,* or limbic lobe, components and *lateral,* or cortical and basal ganglia, components. In his view, the medial part plays a major role in modulating the hypothalamus, whose output pathway is the medial forebrain bundle, and the lateral (cortical–basal ganglia) components, whose output pathway is the internal capsule–lateral forebrain bundle component. As Swanson (1999) concisely states, "Perhaps the fundamental dichotomy proposed by workers throughout the history of neuroscience may ultimately be reduced in general terms to the lateral forebrain (cognitive) system, and the medial forebrain (visceral) system, which clearly must themselves be integrated."

Can We "Reverse Engineer" an Orgasm?

Wikipedia, the online encyclopedia, defines *reverse engineering* as "the process of taking something . . . [e.g., a device] . . . apart and analyzing its working in detail, usually with the intention to construct a new device . . . that does the same thing without actually copying anything from the original." If the device is a brain that produces an orgasm, is it possible to analyze the division of labor among the various brain components and then, conceptually, build a brain that produces an orgasm? In other words, having identified brain components that are involved in orgasm (e.g., nucleus accumbens, cingulate cortex, insular cortex, amygdala, hippocampus, PVN), and knowing something about what these brain components do (based on studies in nonorgasm contexts), can we combine the components to produce an orgasm?

The short answer is no. Or, at best, not yet. This exercise is typical of what many behavioral neuroscientists try to do. They try to understand what various brain components do and how they are orchestrated to generate behavior patterns—eating, drinking, fighting, fleeing, mating, learning, and so forth.

The following discussion gives a glimmering of how daunting the task is. Perhaps the easiest brain component with which to start our reverse-engineering project is the hypothalamic paraventricular nucleus. At least we know that the PVN neurons produce oxytocin, that they become activated during orgasm, that a consequence of their activation is the release of oxytocin into the bloodstream from the posterior pituitary gland, that the uterus responds to the oxytocin surge by contracting vigorously, and that women have said that the vigorous contractions intensify the pleasure of orgasm.

Paraventricular Nucleus of the Hypothalamus

The PVN also plays an essential role in ejaculation. In rats, these neurons are stimulated by dopamine (at D4 receptors) and inhibited by opioid peptides (Melis, Succu, et al., 2005). Axons from neurons in the PVN project to the posterior pituitary, to release oxytocin from there into the bloodstream. Other oxytocin-containing axons from these neurons project down through the spinal cord and synapse on neurons at the lumbar to sacral (lumbosacral) level that control erection and ejaculation, as shown by neuroanatomical tracer (molecular labeling) methods (Veronneau-Longueville et al., 1999). Thus, oxytocin serves both as a hormone in orgasm in women and men and as a neurotransmitter in orgasm in men (Argiolas & Melis, 2005).

We do not know whether the activation of these neurons itself reaches cognitive awareness or in what other ways these neurons might contribute to the perception of orgasm. Regarding other functions of the PVN, one of the authors (Komisaruk) has always been intrigued by a physiological-cognitive event that occurred repeatedly and reliably almost every time his wife nursed their first son. After the infant had been suckling for about one to two minutes, three events would invariably occur in a constant sequence. First, his wife would ask him to get her a glass of water, because suddenly an intense thirst would come over her. Second, she would say something like, "There's that dark curtain or cloud descending over me again, with the brief, transient vague feeling of sadness or depression, hard to describe." And third, within 15 to 20 seconds, almost like clockwork, she would say, "My milk just started to flow."

It takes about 15 to 20 seconds from the time oxytocin is released into the bloodstream from the posterior pituitary until milk is ejected. Therefore, the sudden thirst and the "dark curtain" apparently corresponded in time with the activation of the PVN. On the basis of studies in animals, we know this nucleus controls drinking behavior (Gutman, Jones & Ciriello, 1988). Furthermore, the physiological adaptiveness of this link between secreting milk and drinking is evident (i.e., for the mother to replenish the bodily fluid lost in the milk). But the "dark descending curtain and vague feeling of depression" is a mystery. The intriguing implication, however, is that neurons of the PVN, when activated, can generate cognitive awareness—sudden thirst and hard-to-describe sudden and transient mood change accompanied by correlated visual imagery. If this is difficult to understand, then the role of other brain regions is even more difficult.

Nucleus Accumbens

The nucleus accumbens is activated on orgasm in women, as seen in the fMRI data. This nucleus, located in the front end of the limbic system, receives a significant dopaminergic input from neurons in the ventral tegmental area in the mesencephalic-diencephalic junction region. Holstege et al. (2003) found that this region is activated during orgasm in men. Thus, the same dopaminergic system is evidently activated during orgasm in women and men, and the observed differences in brain regions activated are most likely two sides of the same—mesolimbic dopaminergic system—coin. What appear to be sex differences may well be due just to differences in methodology, since Komisaruk, Whipple, Crawford, et al.'s data are derived from fMRI while Holstege et al.'s are derived from PET, with their attendant differences in sensitivity and resolving power.

Dopaminergic blockade attenuates orgasmic intensity, and dopaminergic stimulation intensifies it. The nucleus accumbens is activated during a nicotine or cocaine "rush," which has been described as an orgasmic feeling . According to Schultz (2000), in rats, "neurons in the nucleus accumbens show activity that is sustained for several seconds when the subject approaches a lever that can be pressed to receive a cocaine injection and stops rapidly when the lever is pressed." And if electrodes are implanted into the nucleus accumbens, rats will bar-press vigorously to obtain electrical stimulation through these electrodes. Thus, it is intriguing that activation of the nucleus accumbens seems to generate not only pleasurable feelings but also their anticipation. Beyond these elements of orgasm—which are difficult enough to understand, let alone design an analog for—the problem gets even more complex.

Cingulate Cortex

At orgasm there are approximate doublings of heart rate, blood pressure, and pupil diameter, all of which indicate a net intense activation of the sympathetic division of the autonomic nervous system. This may involve the cingulate cortex, which becomes activated not only by orgasm (Komisaruk, Whipple, Crawford, et al., 2004) but also by painful stimulation (Casey, Morrow, et al., 2001), both of which conditions also activate the sympathetic division, which increases heart rate, blood pressure, and other "stress" responses.

Electrical stimulation of the cingulate cortex produces increases in respiration and cardiovascular function in animals and humans (Lofving, 1961; Hoff, Kell & Carroll, 1963), indicating its link to autonomic outflow from the sympathetic division. Furthermore, painful stimulation applied to the skin activates the anterior cingulate cortex, as measured by PET (Talbot et al., 1991; Coghill et al., 1994). Conversely, lesions to the anterior cingulate cortex (or to the insular cortex) in humans alter the affective (emotional) responses to pain—the "suffering" (Coghill et al., 1994) or "unpleasant emotional component" of pain (A. K. P. Jones et al., 1991). For example, as reported by Vogt (1999), "Patients with lesions in the anterior cingulate cortex report that they can still identify where the source of noxious sensory stimulation is on the body, but they no longer mind it . . . PET studies show reduced [activity] in anterior cingulate cortex in major depression suggesting a disruption of function in this region." Vogt characterized the anterior cingulate cortex as being involved in "affect regulation."

In another PET study, the cingulate cortex responded not only to painful stimulation but to pleasurable stimulation, consistent with the finding that it becomes active during orgasm (Komisaruk, Whipple, Crawford, et al., 2004). Francis et al. (1999) commented that "the fact that activation of parts of the cingulate cortex, including the subgenual cingulate, was found in at least some subjects . . . [in response to] . . . the pleasant touch, taste and olfactory stimuli suggests that this region is not involved only in the processing of affectively negative stimuli such as pain and aversive taste."

The anterior cingulate cortex is evidently involved in responding with autonomic sympathetic division output not only to direct physiological stimulation such as pain and orgasm but also to more cognitive forms of stress. Critchley et al. (2003) described three patients with damage to the cingulate cortex.

> [The patients had] blunted cardiovascular responses to effortful cognitive task performance (mental stress, overt rapid serial subtraction of 7s) . . . In healthy individuals, the mental stress task elicits significant increases in systolic blood pressure (17 millimeters of mercury) and heart rate (7 beats per minute), which were not observed in these patients . . . anterior cingulate cortex-lesioned patients also had abnormalities in a number of "low level" sympathetic responses, for example to orthostatic challenge [e.g., normal increase in blood pressure on standing up suddenly from a supine position], suggesting that anterior cingulate cortex activity is closely linked to homeostatic autonomic mechanisms.

The patients' performance on cognitive tasks was normal. Critchley et al. concluded that "the observation that patients with anterior cingulate cortex damage fail to generate contextually appropriate autonomic (notably sympathetic) outputs corroborates the anatomical neuroimaging data by providing evidence that this region is *necessary* for the appropriate generation of autonomic arousal during effortful cognitive and physical work."

Other authors postulated a similar, more extended cognitive role for the cingulate cortex (Bush, Luu & Posner, 2000). They pointed out that "in humans, lesions of anterior cingulate cortex for the treatment of affective disorders produce striking personality changes, including lack of distress and emotional lability . . .

[The lesions produce] inattention and akinetic [immobile] states." They proposed a comprehensive, integrative role for the anterior cingulate cortex that includes "specific processing . . . for sensory, motor, cognitive, and emotional information, and on the output side, an influence on . . . activity in other brain regions and [modulation of] cognitive, motor, endocrine and visceral responses."

Thus, the cingulate cortex seems to possess the dual capability of mediating the intense feelings of orgasm while stimulating the autonomic activity characteristic of orgasm. This would seem to place the cingulate cortex in a central role in the generation of orgasm. It does not seem too great a stretch to include the various properties of orgasm in the various descriptions of the cingulate cortex given above.

Insular Cortex (Insula)

The insular cortex, activated during orgasm, mediates visceral feelings, especially pleasure and pain. Using electrical stimulation in humans, Penfield and Faulk (1955) found that stimulation in the anterior insular cortex produced predominantly visceral sensations.

In PET studies, Small et al. (2001) found that when subjects ate chocolate while highly motivated, and rated the chocolate as very pleasant, the insula was one of the areas activated. This suggests that the insular cortex is one of the brain regions where reward value is represented—at least of food, but probably more generalized reward. The primary neural pathway for taste sensation to reach the brain is via the uppermost region of the nucleus of the solitary tract, whose lowermost region receives vaginocervical input. The taste pathway projects from the NTS to the insular cortex, so in some way the insula evidently mediates pleasurable taste and feelings.

However, Coghill et al. (1994), also using PET scans, found that "noxious heat, but not vibrotactile (a buzzer applied gently to the skin) stimulation, was found to activate a region of the contralateral anterior insular cortex . . . This was the only brain region in which we observed significantly greater blood flow during pain than during vibrotactile stimulation." Others pointed out that the neural pathway by which the pain stimulation ascends from the spinal cord is the spinothalamic pathway to the posterior thalamic nuclei, and then via the insula to the amygdala and the perirhinal cortex (Casey, Minoshima, et al., 1994). These authors postulated that this system underlies the "mnemonic [memory], motivational and affective components of pain." And conversely, lesions to the insular cortex (or anterior cingulate cortex) in humans have been reported to alter affective responses to pain.

A more cognitive aversive component—sadness, generated by recall during a PET scan—was associated with an activation in the vicinity of the anterior insular cortex that exceeded the activation produced by film-generated sadness or recall-generated happiness (Reiman et al., 1997). This finding suggested to the investigators that this region participates in the emotional response to potentially distressing cognitive or interoceptive (visceral, gut) sensory stimuli.

It is tempting to speculate that insular cortex activity could be affected by vaginocervical stimulation, not only during orgasm but also during childbirth, and as such may be involved in the occurrence of postpartum depression.

Taken together, the findings described here suggest that the insular cortex may play a significant role in mediating the intense feelings generated by orgasm.

Hippocampus and Amygdala

Activation of both the amygdala and the hippocampus occurs during orgasm. And both the amygdala and the hippocampus are susceptible to temporal lobe, or psychomotor, epilepsy. The seizures start with an aura that begins as hallucinatory sensations, such as olfactory or gustatory illusions (Nauta, 1972), and they may generate orgasmic feelings. Furthermore, lesions of the amygdala can produce "hypersexuality" in many animal species, often directed to inappropriate objects, a condition termed "psychic blindness" (*Seelenblindheit*).

The amygdala thus seems to mediate the feelings of orgasm and the recognition and integration of sexual stimuli. In a way, this is consistent with the findings in a brain imaging study in which watching emotion-laden movies activated the amygdala and the hippocampal formation, "suggesting that these regions participate in the emotional response to certain exteroceptive sensory stimuli" (Reiman et al., 1997).

The amygdala is reciprocally connected with the hippocampus. A major function of the hippocampus is short-term memory. After a lesion in the hippocampus, the patient typically loses short-term memory but retains long-term memory: "As he reads toward the bottom of a printed page, he may already have lost all recollection of what he just read at the top. And as he speaks to a new acquaintance, he may interrupt the conversation to ask a second time for the name and the purpose of his visitor, whom he suddenly does not know" (Nauta & Feirtag, 1986).

Since the hippocampus is activated during orgasm even when the orgasm is generated by thinking, it is likely that the hippocampus plays a significant role in mediating cognitive aspects of orgasm.

The amygdala probably also mediates a cognitive component of orgasm, a conclusion consistent with that of Nauta and Feirtag (1986): "The amygdala directs its cortical projections to the inferior temporal cortex and to the frontal cortex (specifically the orbital surface of the frontal cortex). It projects, therefore, to the parts of the neocortex in which the final stages in the cascade of sensory data toward the limbic system are embodied. Evidently the amygdala screens its neocortical input. Perhaps, then, it intervenes in ideation and cognition . . . It is as if the amygdala were participating in the brain's appreciation of the world."

These brain components that are activated during orgasm are known to be involved in some of the processes that comprise orgasm, such as cardiovascular, visceromotor, hormonal, hedonic, mnemonic, and other cognitive processes. Based on the functions of these brain components in other, nonorgasm contexts, we have some idea of how they may be orchestrated and integrated to generate orgasm. But we are still a long way from the reverse engineering of orgasm.

Some Brain Circuits Underlying the Orgasmic Response

Swanson (1999) provided a very concise and helpful summary of available pathways by which sensory stimulation can access the parts of the brain known to be involved in sexual response and orgasm, how those brain regions interconnect, and their main output pathways to the visceral activity that characterizes orgasm—heart rate, blood pressure, vaginal lubrication, ejaculation, and so forth. The following is a brief summary of these communication channels, based on Swanson's description (1999). It outlines some basic connection pathways among the various brain components that are reported to become activated during sexual arousal and orgasm.

The subicular complex (not the hippocampus proper) projects through the hippocampus to the mammillary bodies of the hypothalamus, the anterior thalamic nuclei, and the entorhinal cortex. The anterior thalamic nuclei and the cingulate gyrus project back to the subicular complex. They also receive input from the mammillary bodies, which in turn receive input from the rest of the hypothalamus and the septum. This is a modernized version of the Papez circuit, based on more recent neuroanatomical evidence.

The subicular complex receives input from the neocortex and provides output to it, and thus is an important mediator between visceral and cognitive functions.

The hippocampus proper sends projections to the lateral septal nucleus, the nucleus accumbens, and parts of the prefrontal cortex. The lateral septal nucleus sends projections to the medial septal nucleus, which projects back to the hippocampus. The septal nuclei also receive and send activity from and to the hypothalamus and lower brainstem.

The cingulate gyrus also projects to neocortical association areas, to the striatum, and to the gray matter of the pons, which in turn projects to the cerebellum. This is a significant pathway by which the "emotional" brain system can influence the (extrapyramidal somatomotor) "muscular movement control" brain system.

Recall that fMRI and PET studies have shown that at orgasm there is significant activation of most of these interconnected brain regions—hippocampus, cingulate cortex, striatum, hypothalamus, and cerebellum. There is also a major communication channel between the hippocampal formation and the amygdala.

The amygdala has four main divisions: (1) the corticomedial division, which receives olfactory input; (2) the basolateral division, which has input from, and sends output to, the neocortex; (3) the central division, which projects to the hypothalamus and parasympathetic nuclei of the brainstem; and (4) the basomedial division, which projects to the mediodorsal nucleus of the thalamus, which in turn projects to the prefrontal neocortex and then back to the basomedial division. The prefrontal neocortex projects to and modulates the sensory systems. The amygdala also projects to the septum and the ventromedial hypothalamus, which in turn project to the brainstem reticular formation, central gray, and autonomic nuclei. During orgasm, the amygdala, reticular formation, and central gray regions are also activated.

Together, these hippocampal and amygdala systems provide the communication channels between the emotional and cognitive brain divisions, integrating visceral, somatic motor, cognitive, and emotional processes, all of which comprise orgasm.

Another sensory modality—pain—is conveyed from the body via lamina I (the outer layer) of the spinal cord to the posterior portion of the ventromedial thalamic nucleus, which projects to the anterior cingulate cortex and then to limbic structures such as the amygdaloid complex and the perirhinal cortex (Coghill et al., 1994).

Olfactory (smell) sense gains access to the pyriform cortex overlying the amygdala and projects to the corticomedial division of the amygdala, from which it projects to at least the hypothalamus.

Visceral sensory activity gains access via the NTS (which is situated in the medulla oblongata at the base of the brainstem, just above the spinal cord), receives information from the internal genital structures in women, the gastrointestinal tract (including taste sense), respiratory system, and cardiac system, and projects to the hypothalamus via a "visceral sensory lemniscus" (a "ribbon" of nerve fibers) (Nauta & Feirtag, 1986). Studies with fMRI found that electrical stimulation of the vagus nerve, a major visceral sensory projection pathway to the brain, activates the cerebellum and thalamus (Chae et al., 2003).

Nauta and Feirtag (1986) provided a succinct answer to the question of how sensory activity accesses this system:

> All of the senses represented in the neocortex—vision, hearing, and the somatic sense—direct part of their traffic toward either or both of two cortical districts: the frontal association cortex and the inferior temporal association cortex. These two regions are connected via a massive fiber bundle—the uncinate fasciculus. In turn, the inferior temporal cortex projects to the entorhinal area, which via the subicular complex is the cortical gateway to the hippocampus. The orbital part of the frontal cortex—alone in the neocortex—projects to the hypothalamus . . . it has uninterrupted access to the visceral, endocrine, and affective mechanisms of the hypothalamic continuum. No other part of the neocortex has access so direct. What, then, is the frontal cortex? . . . It is notable for its lack of primary sensory fields. It is entirely association cortex.

Pain and Pleasure: Strange Brainfellows?

The complexity of reverse engineering of the orgasmic process is clear. Adding to that complexity is the curious observation that sex and pain seem to be linked. Indeed, during sexual intercourse or masturbation leading to orgasm, facial expressions seem to convey pain or suffering. This phenomenon leads us to the studies of brain regions that are activated in response to both painful stimulation and genital stimulation. A. K. P. Jones et al. (1991) found that the anterior cingulate region was activated when painful stimulation was applied in humans, and they proposed that activation of this brain region is the sole representation of the "suffering" component of pain. We also know that the anterior cingulate region is activated when women experience orgasm. While the anterior cingulate cortex is a large and probably diverse brain region, and we do not yet have information on whether pain and orgasm activate the identical neurons in that region, we are tempted to make the following speculation. Perhaps there is a region of the anterior cingulate cortex that generates the *emotion* (which literally means "motion outward") of both pain and orgasm, as distinct from the *feelings* of pain and orgasm, which are diametrically opposite. There must be different neurons that generate the aversive feeling of pain as distinct from the pleasurable feeling of orgasm, but perhaps both sets of neurons project to the same region of the anterior cingulate cortex that generates the common elements of motor expression.

The main trajectory that conveys the sensory signals for pain and for orgasm—the spinothalamic tract—passes through the spinal cord to the brain. It produces intense arousal and activates perhaps a common motor control pathway—via the anterior cingulate cortex. A critical question is, where do these two neural systems—the pain and the pleasure pathways—diverge; where are the neurons that produce the *feeling* of pleasure, as opposed to the *feeling* of pain?

Francis et al. (1999) addressed this issue:

> At the neuronal level, there must be different populations of neurons concerned with reward and punishment, and indeed this has been shown to be the case in for example the primate orbitofrontal cortex for visual, taste, and olfactory stimuli . . . While pleasant touch (velvet), pleasant taste (sugar) and pleasant odor (vanillin) all activated the orbitofrontal cortex, they did so in different regions of that cortex . . . The primary somatosensory cortex responded more to the neutral intense stimulus (a textured piece of wood) than to a pleasant but soft tactile stimulus (velvet). In contrast, the orbitofrontal cortex responded to a greater degree to the pleasant touch stimulus than to the neutral touch.

The question we raised above, about the difference between neurons that produce the feeling of pain versus the feeling of pleasure, leads us to the ultimate question in neuroscience: which of our brain's neurons produce the various qualities of sensations, perceptions, and feelings that we experience? And why does pain feel so bad and pleasure feel so good? An answer to this is the holy grail of neuroscience, and explorations of orgasm are providing some clues, as we continue our attempts to understand orgasm itself.

24

Consciousness and Orgasm

We close this book with a question. Which brain region or regions produce the voluptuous sensation of orgasm? It is the kind of question that scientists ask all the time. We try to figure out how things work, what the workings of one area of the body or the brain can teach us about another, and how we can improve functioning when disorder occurs.

Where in the brain orgasm is produced is a fundamental and unanswered (but, we hope, not unanswerable!) question. If orgasm is a phenomenon of the brain that is more than the sum of the reafferent sensory activity generated from the smooth (involuntary) and striated (voluntary) muscles activated during orgasm—and we believe it probably is "more" than this—we are led inexorably to the question of which neurons generate our experience of pleasure and how they do so. The answer must lie in a realm requiring conceptualization beyond imaging technologies such as PET and fMRI.

We have used terms such as *pleasure* and *pain* loosely throughout the book, as if we know what they mean. Actually, we assume that everyone knows the difference between pleasure and pain, even if none of us can ever really know what anyone else experiences. The issue that this raises is, in its essence, what is consciousness and how is it produced? How can mere chemistry of nerve cells produce such powerful realities as pain or pleasure, hunger and thirst, hot or cold, red or green, a dream, an idea?

Trying to provide an explanation of how neurons can produce consciousness, including the various feelings involved in orgasm, is reminiscent of the following anecdote, in the sense that perhaps we can only get at the question by analogy (at least at present). Albert Einstein related this story when asked to explain his theory of relativity (Esar, 1952):

> I was once walking in the country with a blind friend of mine. The day was hot and I said that I would enjoy a nice cool drink of milk.
>
> "Milk?" asked my friend. "Drink I understand, but what is milk?"
>
> "A white liquid," I explained.
>
> "Liquid I understand, but what is white?"
>
> "The color of a swan's feathers."
>
> "Feathers I understand, but what is a swan?"
>
> "A bird with a crooked neck."
>
> "Neck I understand, but what is crooked?"
>
> I gently took his arm and straightened it. "That's straight," I said. Then I bent it at the elbow. "And this is crooked."
>
> "Oh," exclaimed the blind man, "now I understand what you mean by milk."

At least the question can be narrowed down a bit. While we believe that consciousness is produced by our neurons, we also know that not all the neurons in our body produce awareness. For example, in persons who suffer a disconnection of the spinal cord from the brain, pinching a toe will still elicit a leg withdrawal reflex, but the person does not feel the pinch. The spinal cord neurons are activated by the pinch, but their activity does not produce awareness even though they are part of the person's body.

It is generally believed that consciousness is produced by the activity of neurons in the cerebral cortex. This is probably true. However, there is evidence that noncortical neuronal activity can produce some form of consciousness, as in the case of "blindsight," in which persons who are clinically blind due to visual cortical damage are nevertheless able to make correct visual discriminations, even though they report that they cannot see (Cowey, 2004).

Furthermore, our cortical neurons may be active without producing awareness. For example, one of our colleagues, David Summers, performed an experiment in which he showed research subjects a lighted disc inside a lighted ring, and the disc and surrounding ring could be illuminated separately. There was an eye fixation point at the very center of the combined disc-and-ring pattern. He recorded fMRI activity in the visual cortex of the participants and found, not surprisingly, that when they fixed their gaze on the central fixation point, the lighted ring activated a region of the visual cortex that concentrically surrounded the region activated by the lighted disc. Then he showed both the disc and the ring simultaneously, asking participants to "pay attention" to only the disc or only the ring while continuing to fixate on the central point. As they did so, different parts of the visual cortex were activated when they paid attention to the disc as opposed to the ring, and vice versa. Thus, even though the physical light information was identical, the visual cortex response differed, depending on which of the stimuli the participants paid attention to. Clearly, there must be something different about the activity of neurons that give rise to awareness versus those that don't, at any particular moment in time.

An interesting finding was reported by Benjamin Libet (1999), who found that conscious intention to move, say, a finger, actually *followed* the onset of electrophysiological activity associated specifically with the volitional process. This raises a different, extraordinarily sensitive issue as to the neurophysiological basis of "free will" and the question of the nature of "I." Libet's finding raises the extremely discomforting notion that my neurons make the decision before "I" do. It is the question of whether "I" can ever really decide to do anything, or whether free will is but an illusion produced by the prior activity of the neurons in the brain. Stated another way, can "I" control the activity of my neurons, or do they control "me"? Incidentally, what Libet does not address is, how is the activity of neurons in the brain transduced into conscious experience? In other words, what is the nature of conscious awareness, and how do the neurons that produce it do so? And, while we are at it, what is the minimum number of neurons one would need to produce a "bit" of awareness, and what would they have to do to produce it? To our knowledge, no one throughout recorded history has solved these fundamental questions.

Perhaps one way of trying to understand what consciousness is, is by analogy, much like Einstein's trying to explain "milk" to his friend. The bottom line in the following argument is that consciousness is a reality in a dimension beyond the four dimensions that we think we know and understand.

We believe that we understand the three dimensions of height, width, and depth. What about the fourth dimension, time. Do we really understand it? We cannot synthesize it or put it in a bottle. If we try to imagine what existed before the "big bang" that many physicists think (or, more recently, thought) was the beginning of the universe, the question arises, if there was nothingness, was there time? Could time exist if matter didn't exist? And what about "nothingness"? If there was nothing before the universe existed, was that nothing very, very small or very, very large? Before the universe existed, was there an enormous emptiness or something infinitely small. If the latter, what was "outside" it? And what is outside the universe today?

The point of this mental exercise is to illustrate that it is easy to see that something as commonplace and familiar to us as "time" and "nothing" is quite mysterious and hard to grasp. So if time is a dimension, then it is a dimension that has mysterious qualities as well as commonplace ones.

Perhaps we can apply the same view to consciousness. It can make itself all too evident, as when we experience pain. And it doesn't help if we tell ourselves that pain is just the result of chemical reactions in some cells that happen to be in our brain.

So what if consciousness is a projection of the activity of our neurons into a fifth dimension in which we are embedded at all times? We know it from experiencing it when we are awake. (Why does it disappear when we sleep and reappear in the form of our dreams, which are equally real—just another form of consciousness?) We are just as aware of this fifth dimension of consciousness as we are of the fourth dimension of time, but we cannot bottle either of them. They just have different properties than our more comfortable, graspable three dimensions. But having uniquely different properties is precisely what qualifies something as having a different dimension, just as depth has a property that is different from time.

The thought of consciousness as a fifth dimension is annoying, because it purports to explain something that we really don't understand. But there are many physicists who find it acceptable to contemplate ten or more dimensions. Physicists have the advantage of accepting ideas on the basis that they follow logically and lawfully from accepted (mathematical) rules. One example of this is Einstein's $E = mc^2$. We all have a reasonably comfortable grasp of the concepts of E (energy) and m (mass). If we stretch our imagination, we can grasp the concept of c (the speed of light). But there is no physical concept for the speed of light *squared*! It is a mathematical construct derived from the rules of mathematics. Yet it is a useful descriptor of the relationship between mass and the energy encapsulated in the mass. Using the same basic mathematical methodological constraints and implications, physicists have convinced themselves that there are more than four dimensions.

One can criticize this discussion by saying that this is not science but faith. Faith provides a comforting set of explanations for great questions and useful concepts for complex ideas such as "soul." The great difference between science and faith is that science provides means to *disprove* an idea. In science, we set up tests to show that our idea is wrong. We actively test our theories and try to disprove our hypotheses. And science advances by failed attempts to disprove some nice idea, thereby leading to gradual acceptance of an idea that withstands so many challenges. Nothing is ever truly proven, but some ideas have so much evidence on their behalf, and have been tested so thoroughly, that they seem to be real. By contrast with science, faith provides no means by which to disprove its concepts and, indeed, that is not its mission.

The idea of consciousness as involving an additional dimension is only science if we have a way of disproving the existence of more dimensions. At present we cannot provide a model of how neural activity could be projected onto another dimension that we can somehow simplify into useful everyday activities, such as to go on a green light and stop on a red light. We don't understand how our brain activity—the activity of a hundred billion neurons, with each neuron having as many as a thousand or more synapses—produces, or could produce, this fifth dimension of which we speak. As scientists, we are stuck somewhere between faith and science. We have faith that eventually it will be possible to elucidate the scientific basis for conscious experience, but despite consciousness being so tantalizingly with us, it is exasperatingly elusive. We still have not defined the crucial hypothesis that would allow us to test this fifth dimension and its inhabitant, the feelings of orgasm.

Glossary

adrenergic. Describing a neuron that produces or responds to the neurotransmitter noradrenaline (norepinephrine)

afferent (sensory). Describing incoming neuronal activity to a component of the nervous system

agonist. A substance that stimulates a neuronal receptor for a neurotransmitter; a positive modulator. Contrast **antagonist.**

anorgasmia. Absence of orgasm

antagonist. A substance that blocks the action of a neurologically active substance such as a neurotransmitter, neuromodulator, or hormone. Contrast **agonist.**

atrophy. Tissue wasting

autonomic (involuntary) nervous system. The division of the nervous system that controls functions typically not under voluntary control, such as heart rate, blood pressure, gastrointestinal activity, sweating, salivation, and pupil dilation

axon. Elongated fiber that conveys neural impulses from the neuron cell body toward other neurons

capacitation. The biochemical process that prepares and enables sperm to fertilize an ovum

cell body. The component of a neuron that contains the cell nucleus essential to the neuron's viability

central nervous system. The spinal cord and brain

cholinergic. Describing a neuron that produces or responds to the neurotransmitter acetylcholine

copulation. Mating (of animals, including humans)

corpora cavernosa. A pair of sponge-like, rod-shaped tissues along the length of the penis, containing "sinuses" that become engorged with blood, resulting in erection

cross-over study. A research design in which each research subject first receives either the experimental treatment or the control (placebo) treatment, and subsequently the treatment not previously received. By the end of the study, each subject has received both treatments.

disinhibition. Discontinuation (sometimes by active inhibition) of an ongoing inhibitory process, resulting in a net activation

dopaminergic. Describing a neuron that produces or responds to the neurotransmitter dopamine

double-blind study. A research design in which neither the research subjects nor the researchers know which treatment—experimental or control—is being administered. The "code" is revealed only at the end of the study.

dyspareunia. Discomfort or pain on vaginal stimulation

dystrophy. A degenerative disorder in which muscles weaken and atrophy

female ejaculation. Expulsion of about 3 to 5 milliliters (5 mL = 1 teaspoon) of fluid, different in chemical composition from urine, from the urethra in women

fMRI (functional magnetic resonance imaging). A method of generating images of brain activity in three dimensions based on local changes in the activity of neurons. The neuronal activity changes the local oxygen content of the blood supplying the neurons, which changes the magnetic properties of the iron (hemoglobin) in the blood, producing local perturbations in a strong, imposed, magnetic field; this enables the computerized generation of images that represent the neuronal activity.

GABAergic. Describing a neuron that produces or responds to the neurotransmitter GABA (gamma-aminobutyric acid)

G spot (Gräfenberg spot). A "distinct erotogenic zone" (Gräfenberg, 1950) in women in the anterior vaginal wall surrounding the urethra that swells on sexual stimulation. It can be felt along the course of the urethra by applying finger pressure firmly against the anterior wall of the vagina, about half-way between the back of the pubic bone and the cervix. Named by Perry and Whipple (1981).

half-life. (1) The amount of time required for a radioactive compound to lose half its radioactivity. (2) The amount of time required for the amount of a substance or an activity to decrease by half.

intromission. Insertion of the penis into the vagina or anus

involuntary (smooth) muscle. Muscle that typically is not under voluntary control, such as muscles of the stomach, uterus, and blood vessels. Contrast **voluntary (striated) muscle.**

involuntary (autonomic) nervous system. Neurons that are typically not under voluntary control, such as those that control the stomach, uterus, blood vessels, heart, sweating, and salivation. Contrast **voluntary nervous system.**

ligand. A chemical substance that binds to a receptor on neurons or other cells

limbic system. A conceptual grouping of brain components that are anatomically and functionally linked and are involved in the coordination and control of emotional and motivated behavior such as mating, feeding, drinking, and aggression

lordosis. A mating posture in the females of some animals, such as the rat; elevation of the rump and vaginal opening to facilitate penile insertion

nerve fiber. Elongated component of a neuron that extends beyond the brain and spinal cord and conveys neural activity to or from the periphery (e.g., skin, muscles, viscera). In humans, individual nerve fibers are typically finer than a human hair and may be several feet in length.

neurohormone. A substance produced by a neuron and released directly into the bloodstream, which conveys it to another site in the body where it acts on a gland or a smooth muscle

neuroleptics. A class of drugs used for the treatment of psychosis (e.g., for schizophrenia)

neuromodulator. A substance produced by a neuron that, when released into the synapse, changes the responsiveness of postsynaptic neurons to stimuli impinging on them, without actually stimulating those postsynaptic neurons to fire

neuropeptide. A peptide (a chain of amino acids) produced by a neuron that, when released into the synapse, can act as a neurotransmitter or a neuromodulator

neurotransmitter. A substance produced by a neuron that, when released into the synapse, stimulates or inhibits postsynaptic neurons, muscles, or glands

nongenital orgasm. Feelings that have the properties of genital orgasm but are elicited by stimulation of, and are experienced in, parts of the body other than the genitals

nucleus of the solitary tract (NTS). An elongated cluster of neurons in the medulla oblongata (in the lower brainstem just above the spinal cord) to which the vagus nerves transmit sensory activity originating in thoracic, abdominal, and some pelvic viscera (e.g., heart, stomach, and cervix)

open-label study. A research design in which the (human) research subjects and the researchers know the identity of the substance being administered for testing. Contrast **double-blind study.**

oxytocinergic. Describing a neuron that produces or responds to the neurotransmitter oxytocin

parasympathetic division. Part of the autonomic nervous system; neurons that emanate from the brain or sacral region of the spinal cord that control "rest and recuperation" functions of the body (e.g., digestion and slowing of the heartbeat). Also controls penile erection. Contrast **sympathetic division.**

paraventricular nucleus of the hypothalamus (PVN). A cluster of neurons in the anterior hypothalamus of the brain that are the major source of oxytocin in the body. When activated (by nipple or genital stimulation and during orgasm), these neurons release oxytocin into the bloodstream via axonal projections to the posterior pituitary and axonal projections into the spinal cord.

perineum. The skin region between the anus and the vulva or scrotum

PET (positron emission tomography). A method of imaging brain activity by measuring regional concentrations of an intravenously injected short-lived radioactive substance. As the neuronal activity in a specific region decreases or increases, the blood flow and thus the local concentration of radioactivity decreases or increases. The locations of regional concentrations of radioactivity are mapped in three dimensions by computer and superimposed on anatomical brain scans, providing information on regional changes in brain activity (e.g., in relation to specific sensory stimulation, cognitive tasks, behavior).

placebo-controlled study. A research design in which an intentionally ineffective substance (e.g., "sugar pill") or treatment is administered to some research subjects and the outcomes are compared with the effects of the experimental substance or treatment on other research subjects in the study

portal vessel. A type of blood vessel characterized by a capillary bed at both ends, in contrast with the more common type that has a capillary bed at one end, the other end opening as a tube directly into the heart, lung, or another blood vessel. For example, the portal vessel system between the hypothalamus and the anterior pituitary gland collects "hypothalamic releasing hormones," peptides released from the axon terminals of hypothalamic neurons, and conveys them a short distance (less than an inch) to individual hormone-producing cells in the anterior pituitary gland.

reafferent. Describing a sensory response from an organ that is triggered by its initial sensory response. For example, sensory activity generated by mechanical stimulation of the uterus releases oxytocin from hypothalamic neurons into the bloodstream, and the oxytocin stimulates contraction of uterine muscle. The uterine sensory activity generated by the uterine contractions is reafferent activity.

receptor. A protein molecule located on, or embedded in, a cell membrane or within a cell (e.g., a neuron, a gland cell) that binds selectively to particular ligands (e.g., neurotransmitters, hormones, drugs) and mediates a cascade of metabolic reactions within the cell

reflex. A stereotypical, relatively consistent, specific response to a specific stimulus, such as the knee-jerk, milk-ejection, or lordosis reflexes

retrograde ejaculation. Ejaculation of seminal fluid "backward" into the urinary bladder instead of out through the urethra. This may occur after prostatectomy, which disrupts the neural coordination of the urogenital sphincter muscles.

reuptake. A process by which neurotransmitters, having been released from a neuron to act on postsynaptic neurons, are taken back into the neuron that released them and made available for reuse

serotonergic. Describing a neuron that produces or responds to the neurotransmitter serotonin

sex steroids. A class of hormones secreted by the ovaries or testes that are differentiated by relatively minor variations on a common chemical ("steroid") structure. Include estrogens, androgens, and progestins.

spinal cord. The neurons in the spine (vertebral column) that convey sensory neural impulses from the periphery (e.g., limbs, trunk, viscera) back to the periphery (i.e., spinal reflexes) and to the brain, and convey motor impulses from the brain to the periphery. The spinal cord is divided into four main sections: *cervical* (neck: between skull and topmost rib), *thoracic* (chest or upper back: the level of the ribs), *lumbar* (lower back: between lowest rib and top of pelvis), and *sacral* (level of the pelvis).

spinal nerves. The sensory and motor nerves connected directly to the spinal cord that convey neural impulses to and from the periphery. A left-right pair of sensory and motor spinal nerves passes between the spinal cord and periphery at the junctions between each two successive vertebrae.

SSRIs (selective serotonin reuptake inhibitors). A class of drugs used for the treatment of depression

sympathetic division. Part of the autonomic nervous system; neurons that emanate from the thoracic and lumbar regions of the spinal cord that control "fight or flight" (mobilization of the body's "physical plant" for action). Activation increases heart rate and blood pressure, dilates the airways of the lungs to increase airflow, and shunts blood to the limb muscles and away from the viscera (e.g., stomach, intestine). The net effect is to increase the supply of oxygenated blood to the muscles involved in strenuous action. Activation also produces ejaculation. Contrast **parasympathetic division.**

tonic. Describing persistent, steady contraction of muscle, as in muscle "tone." For example, standing "at attention" involves tonic contraction of the postural muscles of the back and legs.

vasculogenic. Capable of the formation and growth of blood vessels

vasodilation. Dilation of blood vessels, allowing more blood to flow to the region served by those vessels

voluntary (striated) muscle. Muscle that is controlled "voluntarily," such as limb, trunk, and facial muscles. Under the microscope, these muscles are structurally striated. Contrast **involuntary (smooth) muscle.**

voluntary nervous system. The neurons of the brain and spinal cord, and the related peripheral nerves that control the "voluntary" muscles (e.g., of the limbs, trunk, face). Contrast **involuntary (autonomic) nervous system.**

References

Abramov, L. A. 1976. Sexual life and sexual frigidity among women developing acute myocardial infarction. *Psychosomatic Medicine* 38:418–425.

Adams, D. B., Gold, A. B., & Burt, A. D. 1978. Rise in female sexual activity at ovulation blocked by oral contraceptives. *New England Journal of Medicine* 299:1145–1150.

Adler, N. T., Davis, P. G., & Komisaruk, B. R. 1977. Variation in the size and sensitivity of a genital sensory field in relation to the estrous cycle in rats. *Hormones and Behavior* 9:334–344.

Ahlenius, S., & Larsson, K. 1991. Physiological and pharmacological implications of specific effects by 5HT1A agonists on rat sexual behavior. In *5HT1A Agonists, 5HT3 Antagonists and Benzodiazepines: Their Comparative Behavioral Pharmacology,* ed. R. J. Rodgers & S. J. Cooper. New York: John Wiley.

Aizenberg, D., Gur, S., Zenishlany, L., Graniek, M., Jeczmiery, P., & Weizman, A. 1997. Mianserin, a 5HT2A 12C and alpha 2 antagonist, in the treatment of sexual dysfunction by serotonin reuptake inhibitors. *Clinical Neuropharmacology* 20:210–214.

Aizenberg, D., Modai, I., Landa, A., Gil-Ad, I., & Weizman, A. 2001. Comparison of sexual dysfunction in male schizophrenic patients maintained on treatment with classical antipsychotics versus clozapine. *Journal of Clinical Psychiatry* 62:541–544.

Alex, K. D., Yavanian, G. J., McFarlane, H. G., Pluto, C. P., & Pehek, E. A. 2005. Modulation of dopamine release by striatal 5-HT2C receptors. *Synapse* 55:242–251.

Alexander, G. M., Swerdloff, R. S., Wang, C. W., & Davidson, T. 1997. Androgen behavior correlations in hypogonadal men and eugonadal men. I: Mood and response to auditory sexual stimuli. *Hormones and Behavior* 31:110–119.

Alther, L. 1975. *Kinflicks*. New York: New American Library.

Altschuler, S., Rinaman, L., & Miselis, R. 1992. Viscerotopic representation of the alimentary tract in the dorsal and ventral vagal complexes in the rat. In *Neuroanatomy and Physiology of Abdominal Vagal Afferents*, ed. S. Ritter, R. Ritter & C. Barnes. Ann Arbor, MI: CRC Press.

Alzate, H., & Londono, M. L. 1984. Vaginal erotic sensitivity. *Journal of Sex and Marital Therapy* 10:49–56.

Alzate, H., Useche, B., & Villegas, M. 1989. Heart rate change as evidence for vaginally elicited orgasm and orgasm intensity. *Annals of Sex Research* 2:345–357.

American Diabetes Association. 2001. Diabetes and erectile dysfunction. *Clinical Diabetes* 19:48.

American Psychiatric Association. 1994. *Diagnostic and Statistical Manual of Mental Disorders,* 4th ed. Washington, DC: American Psychiatric Association.

Anastasiadis, A. G., Salomon, L., Ghafar, M. A., Burchardt, M., & Shabsigh, R. 2002. Female sexual dysfunction: state of the art. *Current Urology Reports* 3:484–491.

Andersen, K. V., & Bovim, G. 1997. Impotence and nerve entrapment in long distance amateur cyclists. *Acta Neurologica Scandinavica* 95:233–240.

Antelman, S. M., & Rowland, N. E. 1977. The nigrostriatal dopamine system and enhanced reactivity to environmental stimuli: relevance to psychiatric problems. *Activitas Nervosa Superior (Praha)* 19(4):304–306.

Arango, V., Underwood, M. D., & Mann, H. 1992. Alterations in monoamine receptors in the brain of suicide victims. *Journal of Clinical Psychopharmacology* 12(2, suppl.):8S–12S.

Araujo, A. B., Mohr, B. A., & McKinlay, J. B. 2004. Changes in sexual function in middle-aged and older men: longitudinal data from the Massachusetts Male Aging Study. *Journal of the American Geriatrics Society* 52:1502–1511.

Argiolas, A., & Melis, M. R. 2005. Central control of penile erection: role of the paraventricular nucleus of the hypothalamus. *Progress in Neurobiology* 76:1–21.

Arnott, S., & Nutt, D. 1994. Successful treatment of fluvoxamine induced anorgasmia with cyproheptadine. *British Journal of Psychiatry* 164:838–839.

Arnow, B. A., Desmond, J. E., Banner, L. L., Glover, G. H., Solomon, A., Polan, M. L., Lue, T. F., & Atlas, S. W. 2002. Brain activation and sexual arousal in healthy, heterosexual males. *Brain* 125:1014–1023.

Aron, A., Fisher, H., Mashek, D. J., Strong, G., Li, H., & Brown, L. L. 2005. Reward, motivation, and emotion systems associated with early-stage intense romantic love. *Journal of Neurophysiology* 94:327–337.

Arthurs, O. J., & Boniface, S. 2002. How well do we understand the neural origins of the fMRI BOLD signal? *Trends in Neurosciences* 25:27–31.

Ashton, A. K., & Bennett, R. G. 1999. Sildenafil treatment of serotonin reuptake-inhibitors-induced sexual dysfunction. *Journal of Clinical Psychiatry* 60:194–195.

Ashton, A. K., & Rosen, R. C. 1998. Bupropion as an antidote for serotonin reuptake inhibitor-induced sexual dysfunction. *Journal of Clinical Psychiatry* 59:112–115.

Avis, N. E., Stellato, R., Crawford, S., Johannes, C., & Longcope, C. 2000. Is there an association between menopause status and sexual functioning? *Menopause* 7:297–309.

Bachmann, G. A. 1995. Influence of menopause on sexuality. *International Journal of Fertility and Menopausal Studies (Suppl.)* 40:16–22.

Bachmann, G. A., & Leiblum, S. R. 2004. The impact of hormones on menopausal sexuality: a literature review. *Menopause* 11:120–130.

Baier, D., & Philipp, M. 1994. Effects of antidepressants on sexual function (in German). *Fortschritte der Neurologie-Psychiatrie* 62:14–21.

Baker, R. R., & Bellis, M. A. 1993. Human sperm competition: ejaculate manipulation by females and a function for the female orgasm. *Animal Behaviour* 46:887–909.

Baker, R. R., & Bellis, M. A. 1995. *Human Sperm Competition: Copulation, Masturbation and Infidelity.* London: Chapman and Hall.

Baker, W., Reese, J., & Ito, T. 2006. A randomized, double-blind placebo controlled study to evaluate the ability of a natural nutritional supplement (ArginMax) to enhance the efficacy of Viagra (sildenafil citrate). Unpublished ms.

Baldwin, D. S. 2004. Sexual dysfunction associated with antidepressant drugs. *Expert Opinion on Drug Safety* 3:457–470.

Bancroft, J. 1984. Hormones and human sexual behavior. *Journal of Sex and Marital Therapy* 10:3–21.

Bancroft, J. 1989. *Human Sexuality and Its Problems.* New York: Churchill Livingstone.

Bancroft, J., Danders, D., Davidson, D. W., & Warner, P. 1983. Mood, sexuality, hormones and the menstrual cycle. II: Sexuality and the role of androgens. *Psychosomatic Medicine* 45:509–516.

Bancroft, J., Loftus, J., & Long, J. S. 2003. Distress about sex: a national survey of women in heterosexual relationships. *Archives of Sexual Behavior* 32:193–211.

Barak, Y., Achiron, A., Elizur, A., Gabbay, U., Noy, S., & Sarova-Pinhas, I. 1996. Sexual dysfunction in relapsing-remitting multiple sclerosis: magnetic resonance imaging, clinical, and psychological correlates. *Journal of Psychiatry and Neuroscience* 21:255–258.

Barbach, L. G. 1975. *For Yourself: The Fulfillment of Female Sexuality.* New York: Doubleday.

Barrett, M. 1999. *Sexuality and Multiple Sclerosis.* Toronto: Multiple Sclerosis Society of Canada.

Basson, R. 2001. Female sexual response: the role of drugs in the management of sexual dysfunction. *Obstetrics and Gynecology* 98:350–353.

Basson, R., Althof, S., Davis, S., Fugl-Meyer, K., Goldstein, I., Leiblum, S., Meston, C., Rosen, R., & Wagner, G. 2004. Summary of the recommendations on sexual dysfunctions in women. *Journal of Sexual Medicine* 1:24–34.

Basson, R., Berman, J., Burnett, A., Derogatis, L., Ferguson, D., Fourcroy, J., Goldstein, I., Graziottin, A., Heiman, J., Laan, E., Leiblum, S., Padma-Nathan, H., Rosen, R., Segraves, K., Segraves, R. T., Shabsigh, R., Sipski, M., Wagner, G., & Whipple, B. 2000. Report of the International Consensus Development Conference on female sexual dysfunction: definitions and classifications. *Journal of Urology* 163:888–893.

Basson, R., Leiblum, L., Brotto, L., Derogatis, L., Fourcroy, J., Fugl-Meyer, K., Graziottin, A., Heiman, J. R., Laan, E., Meston, C., Schover, L., van Lankfeld, J., & Weijmar Schultz, W. C. M. 2003. Definitions of women's sexual dysfunction reconsidered: advocating expansion and revision. *Journal of Psychosomatic Obstetrics and Gynecology* 24:221–229.

Basson, R., Weijmar Schultz, W. C. M., Brotto, L. A., Binik, Y. M., Eschenbach, D. A., Laan, E., Utian, W. H., Wesselman, U., van Lankfeld, J., Wyatt, G.,

& Wyatt, L. 2004. *Second International Consultation on Sexual Medicine: Men and Women's Sexual Dysfunction.* Paris: Health Publications.

Bayliss, W. M., & Starling, E. H. 1902. The mechanism of pancreatic secretion. *Journal of Physiology* 28:325–353.

Beach, F. A. 1942. Effects of testosterone propionate upon the copulatory behavior of sexually inexperienced male rats. *Journal of Comparative Psychology* 33:227–247.

Beach, F. 1948. *Hormones and Behavior.* New York: Paul B. Hoeber.

Beach, F. A. 1975. Hormonal modifications of sexually dimorphic behavior. *Psychoneuroendocrinology* 1:3–23.

Bellerose, S. B., & Binik, Y. M. 1993. Body image and sexuality in oophorectomized women. *Archives of Sexual Behavior* 22:435–459.

Bemelmans, B. L., Meuleman, E. J., Doesburg, W. H., Notermans, S. L., & Debruyne, F. M. 1994. Erectile dysfunction in diabetic men: the neurological factor revisited. *Journal of Urology* 151:884–889.

Berard, E. J. J. 1989. The sexuality of spinal cord injured women: physiology and pathophysiology—a review. *Paraplegia* 27:99–112.

Beric, A., & Light, J. K. 1993. Anorgasmia in anterior spinal cord syndrome. *Journal of Neurology, Neurosurgery, and Psychiatry* 56:548–551.

Berkley, K. J., Hotta, H., Robbins, A., & Sato, Y. 1990. Functional properties of afferent fibers supplying reproductive and other pelvic organs in pelvic nerve of female rats. *Journal of Neurophysiology* 63:256–272.

Best, B. 1990. 30–40% of all synapses utilize GABA. www.benbest.com /science/anatmind/anatmd10.html.

Beyer, C., Larsson, K., Peréz-Palacios, G., & Moralí, G. 1973. Androgen structure and male sexual behavior in the castrated rat. *Hormones and Behavior* 4:99–108.

Beyer, C., Morali, G., Larsson, K., & Sodersten, P. 1976. Steroid regulation of sexual behavior. *Journal of Steroid Biochemistry* 7:1171–1176.

Beyer, C., & Rivaud, N. 1973. Differential effect of testosterone and dihydrotestosterone on the sexual behavior of prepubertally castrated male rabbits. *Hormones and Behavior* 4:175–180.

Beyer, C., Vidal, N., & Mijares, A. 1970. Probable role of aromatization in the induction of estrous behavior by androgens in the ovariectomized rabbit. *Endocrinology* 87:1386–1389.

Beyer, C., Yaschine, T., & Mena, F. 1964. Alterations in sexual behaviour induced by temporal lobe lesions in female rabbits. *Boletin de Instituto de Estudios Medicos y Biologicos, Universidad Nacional Autonoma de Mexico* 22:379–386.

Bianca, R., Sansone, G., Cueva-Rolon, R., Gomez, L. E., Ganduglia-Pirovano, M., Beyer, C., Whipple, B., & Komisaruk, B. R. 1994. Evidence that the vagus nerve mediates a response to vaginocervical stimulation after spinal cord transection in the rat. *Society for Neuroscience Abstracts* 20:961.

Bird, V. G., Brackett, N. L., Lynne, C. M., Aballa, T. C., & Ferrell, S. M. 2001. Reflexes and somatic responses as predictors of ejaculation by penile vibratory stimulation in men with spinal cord injury. *Spinal Cord* 39:514–519.

Bjorklund, A., & Lindvall, O. 1984. Dopamine containing systems in the CNS. In *Handbook of Chemical Neuroanatomy: Classical Transmitters in the CNS,* vol. 2, pt. 1, ed. A. Bjorklund & T. Hokfelt. Elsevier: Amsterdam.

Blaicher, W., Gruber, D., Bieglmayer, C., Blaicher, A. M., Knogler, W., & Huber, J. C. 1999. The role of oxytocin in relation to female sexual arousal. *Gynecologic and Obstetric Investigation* 47:125–126.

Blank, J. 1994. Toys: Sex toys. In *Human Sexuality: An Encyclopedia*, ed. V. L. Bullough & B. Bullough. New York: Garland.

Bliesener, N., Albrecht, S., Schwager, A., Weckbecker, K., Lichtermann, D., & Klingmuller, D. 2005. Plasma testosterone and sexual function in men receiving buprenorphine maintenance for opioid dependence. *Journal of Clinical Endocrinology and Metabolism* 90:203–206.

Bloom, F. E. 2001. Neurotransmission and the central nervous system. In *Goodman & Gilman's The Pharmacological Basis of Therapeutics,* ed. 10, ed. J. G. Hardman, L. E. Limbird & A. G. Gilman. New York: McGraw-Hill.

Blumer, D. 1970. Hypersexual episodes in temporal lobe epilepsy. *American Journal of Psychiatry* 126:1099–1106.

Bodnar, R. J., Commons, K., & Pfaff, D. W. 2002. *Central Neural States Relating Sex and Pain.* Baltimore: Johns Hopkins University Press.

Bohlen, J. G., Held, J. P., & Sanderson, M. O. 1980. The male orgasm: pelvic contractions measured by anal probe. *Archives of Sexual Behavior* 9:503–521.

Bohlen, J. G., Held, J. P., Sanderson, M. O., & Ahlgren, A. 1982. The female orgasm: pelvic contractions. *Archives of Sexual Behavior* 11:367–386.

Bonica, J. J. 1967. *Principles and Practices of Obstetric Analgesia and Anesthesia.* Philadelphia: F. A. Davis.

Bonsall, R. W., Rees, H. D., & Michael, R. P. 1989. Identification of radioactivity in cell nuclei from brain, pituitary gland and reproductive tract of male rhesus monkeys after the administration of [3H] testosterone. *Journal of Steroid Biochemistry* 32:599–608.

Boolell, M., Allen, M. J., Ballard, S. A., Gepi-Attee, S., Muirhead, G. J., Naylor, A. M., Osterloh, I. H., & Gingell, C. 1996. Sildenafil: an orally active type 5 cyclic GMP-specific phosphodiesterase inhibitor for the treatment of penile erectile dysfunction. *International Journal of Impotence Research* 8:47–52.

Bornhovd, K., Quante, M., Glauche, V., Bromm, B., Weiller, C., & Buchel, C. 2002. Painful stimuli evoke different stimulus-response functions in the amygdala, prefrontal, insula and somatosensory cortex: a single-trial fMRI study. *American Journal of Gastroenterology* 97:654–661.

Bors, E., & Comarr, A. E. 1960. Neurological disturbances of sexual function with special reference to 529 patients with spinal cord injury. *Urological Survey* 10:191–221.

Bowers, M. B., Van Woert, M., & Davis, L. 1971. Sexual behavior during L-dopa treatment for parkinsonism. *American Journal of Psychiatry* 127:1691–1693.

Bradshaw, H. B., & Berkley, K. J. 2000. Estrous changes in responses of rat gracile nucleus neurons to stimulation of skin and pelvic viscera. *Journal of Neuroscience* 20:7722–7727.

Brannon, E. G., & Rolland, P. D. 2000. Anorgasmia in a patient with bipolar disorder type I treated with gabapentin. *Journal of Clinical Psychopharmacology* 20:379–381.

Braunstein, G. D., Sundwall, D. A., Katz, M., Shifren, J. L., Buster, J. E., Simon, J. A., Bachman, G., Aguirre, O. A., Lucas, J. D., Rodenberg, C., Buch, A., & Watts, N. B. 2005. Safety and efficacy of a testosterone patch for the treatment of hypoactive sexual desire disorder in surgically menopausal women. *Archives of Internal Medicine* 165:1582–1589.

Breiter, H. C., Gollub, R. L., Weisskoff, R. M., Kennedy, D. N., Makris, N., Berke, J. D., Goodman, J. M., Kantor, H. L., Gastfriend, D. R., Riorden, J. P., Mathew, R. T., Rosen, B. R., & Hyman, S. E. 1997. Acute effects of cocaine on human brain activity and emotion. *Neuron* 19:591–611.

Brick, P., & Lunquist, J. 2003. *New Expectations: Sexual Education for Mid and Later Life*. New York: Sexuality Information and Education Council of the United States.

Brindley, G. S. 1983. Physiology of erection and management of paraplegic infertility. In *Male Infertility*, ed. T. B. Hargreave. New York: Springer-Verlag.

Brindley, G. S. 1986. Pilot experiments on the actions of drugs injected into the human corpus cavernosum penis. *British Journal of Pharmacology* 87:495–500.

Bronner, G., Royter, V., Korczyn, A., & Giladi, N. 2004. Sexual dysfunction in Parkinson's disease. *Journal of Sex and Marital Therapy* 30:95–105.

Bruchovsky, N., & Wilson, J. D. 1968. The conversion of testosterone 5α-androstan 17B-al-3-one by rat prostate in vivo and in vitro. *Journal of Biological Chemistry* 243:2012–2021.

Buchman, M. T., & Kellner, R. 1984. Reduction of distress in hyperprolactinemia with bromocriptine. *American Journal of Psychiatry* 6:357–358.

Bucy, P. C., & Kluver, H. 1955. An anatomical investigation of the temporal lobe in the monkey (*Macaca mulatta*). *Journal of Comparative Neurology* 103:151–251.

Buijs, R. M., Geffard, M., Pool, C. W., & Hoorneman, E. M. D. 1984. The dopaminergic innervation of the supraoptic and paraventricular nucleus: a light and electron microscopical study. *Brain Research* 323:65–74.

Burroughs, W. S. 1959. *Naked Lunch*. New York: Grove Press.

Bush, G., Luu, P., & Posner, M. I. 2000. Cognitive and emotional influences in anterior cingulate cortex. *Trends in Cognitive Sciences* 4:215–222.

Caggiula, A. R. 1970. Analysis of the copulation-reward properties of posterior hypothalamic stimulation in male rats. *Journal of Comparative Physiology and Psychology* 70:399–412.

Caggiula, A. R., Herndon, J. G., Jr., Scanlon, R., Greenstone, D., Bradshaw, W., & Sharp, D. 1979. Dissociation of active from immobility components of sexual behavior in female rats by central 6-hydroxydopamine: implications for CA involvement in sexual behavior and sensorimotor responsiveness. *Brain Research* 172:505–520.

Cain, V. S., Johannes, C. B., Avis, N. E., Mohr, B., Schocken, M., Skurnick, J., & Ory, M. 2003. Sexual functioning and practices in a multi-ethnic study of midlife women: baseline results from SWAN. *Journal of Sex Research* 40:266–276.

Calleja, J., Carpizo, R., & Berciano, J. 1988. Orgasmic epilepsy. *Epilepsia* 29:635–639.

Cantor, J. M., Binik, Y. M., & Pfaus, J. G. 1999. Chronic fluoxetine inhibits sexual behavior in the male rat: reversal with oxytocin. *Psychopharmacology* 144:355–362.

Carani, C., Granata, A. R., Rochira, V., Caffagni, G., Aranda, C., Anunez, P., & Maffei, L. E. 2005. Sex steroids and sexual desire in a man with a novel mutation of aromatase gene and hypogonadism. *Psychoneuroendocrinology* 30:413–417.

Carmichael, M. S., Humbert, R., Dixen, J., Palmisano, G., Greenleaf, W., & Davidson, J. M. 1987. Plasma oxytocin increases in the human sexual response. *Journal of Clinical Endocrinology and Metabolism* 64:27–31.

Carmichael, M. S., Warburton, V. L., Dixen, J., & Davidson, J. M. 1994. Relationships among cardiovascular, muscular, and oxytocin responses during human sexual activity. *Archives of Sexual Behavior* 23:59–79.

Carter, C. S., Williams, J. R., Witt, D. M., & Insel, T. R. 1992. Oxytocin and social bonding. *Annals of the New York Academy of Sciences* 652: 204–211.

Caruso, S., Agnello, C., Intelisano, G., Farina, M., DiMari, L., & Cianci, A. 2004. Sexual behavior of women taking low-dose oral contraceptive containing 15 microg ethinylestradiol/60 microg gestodene. *Contraception* 69:237–240.

Casey, K. L., Minoshima, S., Berger, K. L., Koeppe, R. A., Morrow, T. J., & Frey, K. A. 1994. Positron emission tomographic analysis of cerebral structures activated specifically by repetitive noxious heat stimuli. *Journal of Neurophysiology* 71:802–807.

Casey, K. L., Morrow, T. J., Lorenz, J., & Minoshima, S. 2001. Temporal and spatial dynamics of human forebrain activity during heat pain: analysis by positron emission tomography. *Journal of Neurophysiology* 85:951–959.

Centers for Disease Control. 1991. *HIV/AIDS Surveillance Report.* Atlanta: U.S. Department of Health and Human Services.

Chae, J. H., Nahas, Z., Lomarev, M., Denslow, S., Lorberbaum, J. P., Bohning, D. E., & George, M. S. 2003. A review of functional neuroimaging studies of vagus nerve stimulation (VNS). *Journal of Psychiatric Research* 37:443–455.

Chapelle, P. A., Durand, J., & Lacert, P. 1980. Penile erection following complete spinal cord injury in man. *British Journal of Urology* 52:216–219.

Charnetski, C. J., & Brennan, F. X. 2001. *Feeling Good Is Good for You: How Pleasure Can Boost Your Immune System and Lengthen Your Life.* Emmaus, PA: Rodale Press.

Chia, M., & Abrams, R. C. 2005. *The Multi-orgasmic Woman.* New York: Rodale Press.

Chia, M., & Arava, D. A. 1996. *The Multi-orgasmic Man.* San Francisco: HarperSanFrancisco.

Chia, M., Chia, M., Abrams, D., & Abrams, R. C. 2000. *The Multi-orgasmic Couple.* San Francisco: HarperSanFrancisco.

Chiriac, J. 2004. Freud and the "Cocaine Episode." www.freudfile.org/cocaine.html.

Choi, H. K., & Seong, D. H. 1995. Effectiveness for erectile dysfunction after the administration of Korean red ginseng. *Korean Journal of Ginseng Science* 19:17–21.

Chuang, Y.-C., Lin, T.-K., Lui, C.-C., Chen, S.-D., & Chang, C.-S. 2004. Tooth-brushing epilepsy with ictal orgasms. *Seizure* 13:179–182.

Cipolotti, L., Shallice, T., & Dolan, R. J. 2003. Human cingulate cortex and autonomic control: converging neuroimaging and clinical evidence. *Brain* 126:2139–2152.

Clayton, A. H., Zajecka, L., Ferguson, J. M., Filipiak-Reisnier, J. K., Brown, M. T., & Schwartz, G. E. 2003. Lack of sexual dysfunction with the selective noradrenaline reuptake inhibitor reboxetine during treatment for major depressive disorder. *International Clinical Psychopharmacology* 18:151–156.

Clayton, D. O., & Shen, W. W. 1998. Psychotropic drug-induced sexual function disorders: diagnosis, incidence and management. *Drug Safety* 19:299–312.

Coghill, R. C., Talbot, J. D., Evans, A. C., Meyer, E., Gjedde, A., Bushnell, M. C., & Duncan, G. H. 1994. Distributed processing of pain and vibration by the human brain. *Journal of Neuroscience* 14:4095–4108.

Cohen, A. J., & Bartlik, B. 1998. Ginkgo biloba for antidepressant-induced sexual dysfunction. *Journal of Sex and Marital Therapy* 124:139–143.

Cole, T. 1975. Sexuality and physical disabilities. *Archives of Sexual Behavior* 4:389–403.

Collins, J. J., Lin, C. E., Berthoud, H. R., & Papka, R. E. 1999. Vagal afferents from the uterus and cervix provide direct connections to the brainstem. *Cell and Tissue Research* 295:43–54.

Comarr, A. E., & Vigue, M. 1978. Sexual counseling among male and female patients with spinal cord injury and/or cauda equina injury. Parts I and II. *American Journal of Physical Medicine* 57:107–227.

Coolen, L. M., Allard, J., Truitt, W. A., & McKenna, K. E. 2004. Central regulation of ejaculation. *Physiology and Behavior* 83:203–215.

Cooper, A. J., Cernovsky, Z. Z., & Colussi, K. 1993. Some clinical and psychometric characteristics of primary and secondary premature ejaculators. *Journal of Sex and Marital Therapy* 19:276–88.

Cooper, J. R., Bloom, F. E., & Roth, R. H. 2003. *The Biochemical Basis of Neuropharmacology.* New York: Oxford University Press.

Cordoba, O. A., & Chapel, J. L. 1983. Medroxyprogesterone acetate antiandrogen treatment of hypersexuality in a pedophilic sex offender. *American Journal of Psychiatry* 140:1036–1039.

Cowey, A. 2004. The 30th Sir Frederick Bartlett lecture: Fact, artifact, and myth about blindsight. *Quarterly Journal of Experimental Psychology. A, Human Experimental Psychology* 57:577–609.

Critchley, H. D., Mathias, C. J., Josephs, O., O'Doherty, J., Zanini, S., Dewar, B. K., Cipolotti, L., Shallice, T., & Dolan, R. J. 2003. Human cingulate cortex and autonomic control: converging neuroimaging and clinical evidence. *Brain* 126:2139–2152.

Cross, B. A., & Wakerley, J. B. 1977. The neurohypophysis. *International Reviews of Physiology* 16:1–34.

Crouch, N. S., Minto, C. L., Laio, L.-M., Woodhouse, C. R. J., & Creighton, S. M. 2004. Genital sensation after feminizing genitoplasty for congenital adrenal hyperplasia: a pilot study. *BJU International* 93:135–138.

Crowley, T. J., & Simpson, R. 1978. Methadone dose and human sexual behavior. *International Journal of the Addictions* 13:285–295.

Cueva-Rolon, R., Sansone, G., Bianca, R., Gomez, L. E., Beyer, C., Whipple, B., & Komisaruk, B. R. 1996. Vagotomy blocks responses to vaginocervical stimulation in genitospinal-neurectomized rats. *Physiology and Behavior* 60:19–24.

Cunningham, S. T., Steinman, J. L., Whipple, B., Mayer, A. D., & Komisaruk, B. R. 1991. Differential roles of hypogastric and pelvic nerves in the analgesic and motoric effects of vaginocervical stimulation in rats. *Brain Research* 559:337–343.

Cutler, W. B., Zacker, M., McCoy, N., Genovese-Stone, E., & Friedman, E. 2000. Sexual response in women. *Obstetrics and Gynecology* 95(4, suppl. 1):S19.

Cytowic, R. 1998. *The Man Who Tasted Shapes*. Cambridge, MA: MIT Press.

Daniels, G. E., & Tauber, E. S. 1941. A dynamic approach to the study of replacement therapy in cases of castration. *American Journal of Psychiatry* 97:905–918.

DasGupta, R., Kanabar, G., & Fowler, C. 2002. Pudendal somatosensory evoked potentials in women with female sexual dysfunction and multiple sclerosis. *International Journal of Impotence Research* 14:S83.

DasGupta, R., Wiseman, O. J., Kanabar, G., & Fowler, C. J. 2004. Efficacy of sildenafil in the treatment of female sexual dysfunction due to multiple sclerosis. *Journal of Urology* 171:1189–1193.

Davenport, H. W. 1991. Early history of the concept of chemical transmission of the nerve impulse. *Physiologist* 34:129–190.

Davey Smith, G., Frankel, S., & Yarnell, J. 1997. Sex and death: are they related? Findings from the Caerphilly cohort study. *BMJ (Clinical Research ed.)* 315:1641–1644.

Davidson, J. M. 1980. The psychobiology of sexual experience. In *The Psychobiology of Consciousness,* ed. J. M. Davidson & R. J. Davidson. New York: Plenum Press.

Davis, A., Gilbert, K., Misiowiec, P., & Riegel, B. 2003. Perceived effect of testosterone replacement therapy in perimenopausal and postmenopausal women: an Internet pilot study. *Health Care for Women International* 24:831–848.

Davis, K. B. 1929. *Factors in the Sex Life of Twenty-two Hundred Women.* New York: Harpers.

Davis, S. R., Davison, S. L., Donath, S., & Bell, R. J. 2005. Circulating androgen levels and self-reported sexual function in women. *JAMA: Journal of the American Medical Association* 294:91–96.

Davis, S. R., Guay, A. T., Shifren, J. L., & Mazer, N. A. 2004. Endocrine aspects of female sexual dysfunction. *Journal of Sexual Medicine* 1:82–86.

De Amicis, L. A., Goldberg, D. C., LoPiccolo J., Friedman, J., & Davies, L. 1985. Clinical follow-up of couples treated for sexual dysfunction. *Archives of Sexual Behavior* 14:467–489.

Delay, J., Deniker, P., & Harl, M. 1952. Traitement des états d'excitation et d'agitation par une méthode medicamenteuse derivée de l'hibernotherapie [Therapeutic method derived from hiberno-therapy in excitation and agitation states]. *Annales Médico-Psychologiques (Paris)* 110:267–273.

De Leon, G., & Wexler, H. K. 1973. Heroin addiction: its relation to sexual behavior and sexual experience. *Journal of Abnormal Psychology* 81:36–38.

De Lignieres, B. 1993. Transdermal dihydrotestosterone treatment of andropause. *Annals of Medicine* 25:235–241.

Dennerstein, L., Lehert, P., Burger, H., & Dudley, E. 1999. Factors affecting sexual function of women in the mid-life years. *Climacteric* 2:254–262.

Dieckmann, G., Schneider-Joneitz, B., & Schneider, H. 1988. Psychiatric and neuropsychological findings after stereotactic hypothalamotomy, in cases of extreme sexual aggressivity. *Acta Neurochirurgica (Wien) Supplement* 44:163–166.

Ding, Y. Q., Shi, J., Wang, D. S., Xu, J. Q., Li, J. L., & Ju, G. 1999. Primary afferent fibers of the pelvic nerve terminate in the gracile nucleus of the rat. *Neuroscience Letters* 272:211–214.

Dixson, A. F. 1998. *Primate Sexuality: Comparative studies of the Prosimians, Monkeys, Apes and Human Beings*. Oxford: Oxford University Press.

Dixson, A. F., & Herbert, J. 1977. Gonadal hormones and sexual behavior in groups of adult talapoin monkeys (*Miopithecus talapoin*). *Hormones and Behavior* 8:141–154.

DSM-IV-TR. 2000. www.Behavenet.com/capsules/disorders/dsm4TRclassification.htm.

Dunn, K. M., Cherkas, L. F., & Spector, T. D. 2005. Genetic influences on variation in female orgasmic function: a twin study. *Biology Letters* 1:260–263.

Dunn, M. E., & Trost, J. E. 1989. Male multiple orgasms: a descriptive study. *Archives of Sexual Behavior* 18:377–387.

Eisenbach, M. 1995. Sperm changes enabling fertilization in mammals. *Current Opinion in Endocrinology and Diabetes* 2:468–475.

Elliott, H. C. 1969. *Textbook of Neuroanatomy*. Philadelphia: J. B. Lippincott.

Ellison, C. R. 2000. *Women's Sexualities: Generations of Women Share Intimate Secrets of Sexual Self-Acceptance*. Oakland, CA: New Harbinger.

Enzlin, P., Mathieu, C., & Demytteanere, K. 2003. Diabetes and female sexual functioning: a state-of-the-art. *Diabetes Spectrum* 16:256–259.

Erowid. 2005. Heroin withdrawal increases libido. www.erowid.org/chemicals/heroin_faq.shtml.

Erskine, M. S., & Hanrahan, S. B. 1997. Effects of paced mating on c-fos gene expression in the female rat brain. *Journal of Neuroendocrinology* 9:903–912.

Ervin, F. R. 1953. The frontal lobes: a review of the literature. *Diseases of the Nervous System* 14:73–83.

Esar, E. 1952. *The Humor of Humor*. New York: Horizon Press.

Evans, R. W., & Couch, J. R. 2001. Orgasm and migraine. *Headache* 41:512–514.

Everitt, B. J., Fuxe, K., & Hokfelt, T. 1974. Inhibitory role of dopamine and 5-hydroxytryptamine in the sexual behaviour of female rats. *European Journal of Pharmacology* 29:187–191.

Everitt, B. J., & Herbert, J. 1971. The effects of dexamethasone and androgens on sexual receptivity of female rhesus monkeys. *Journal of Endocrinology* 51:575–588.

Everitt, B. J., Herbert, J., & Hamer, J. D. 1972. Sexual receptivity of bilaterally adrenalectomized female rhesus monkeys. *Physiology and Behavior* 8:409–415.

Fadul, C. E., Stommel, E. W., Dragnev, K. H., Eskey, C. J., & Dalmau, J. O. 2005. Focal paraneoplastic limbic encephalitis presenting as orgasmic epilepsy. *Journal of Neurooncology* 72:195–198.

Faerman, I., Jadzinsky, J., & Podolsky, S. 1980. Diabetic neuropathy and sexual dysfunction. In *Clinical Diabetes: Modern Management,* ed. S. Podolsky. New York: Appleton-Century-Crofts.

Faix, A., LaPray, J. F., Callede, O., Maubon, A., & Lanfrey, K. 2002. Magnetic resonance imaging (MRI) of sexual intercourse: second experience in missionary position and initial experience in posterior position. *Journal of Sex and Marital Therapy* 28:63–76.

Farrell, S. A., & Kieser, K. 2000. Sexuality after hysterectomy. *Obstetrics and Gynecology* 95:1045–1050.

Farrington, A. 2005. Female sexual health in midlife and beyond: addressing female sexual distress. *Health and Sexuality* 10:2–16.

Fava, M., & Borofsky, G. F. 1991. Sexual disinhibition during treatment with a benzodiazepine: a case report. *Journal of Psychiatric Medicine* 21:99–104.

Fedele, D., Coscelli, C., Santeusanio, F., Bortolotti, A., Chatenoud, L., Colli, E., Landoni, M., & Parazzini, F. 1998. Erectile dysfunction in diabetic subjects in Italy. *Diabetes Care* 21:1973–1977.

Feiger, A., Kiev, A., Shrivastava, R. K., Wisselink, P. G., & Wilcox, C. S. 1996. Nefazodone versus sertraline in outpatients with major depression: focus on efficacy, tolerability and effect on sexual function and satisfaction. *Journal of Clinical Psychiatry* 57(suppl. 2):53–62.

Feldman, H. A., Goldstein, I., Hatzichristou, D. G., Krane, R. J., & McKinlay, J. B. 1994. Impotence and its medical and psychological correlates: results of the Massachusetts Male Aging Study. *Journal of Urology* 151:54–61.

Feldman, H. A., Johannes, C. B., McKinlay J. B., & Longcope, C. 1998. Low dehydroepiandrosterone sulfate and heart disease in middle-aged men: cross-sectional results from the Massachusetts Male Aging Study. *Annals of Epidemiology* 8:217–228.

Ferguson, D. M., Steidle, C. P., Singh, G. S., Alexander, J. S., Weihmiller, M. K., & Crosby, M. G. 2003. Randomized, placebo-controlled, double-blind, crossover design trial of the efficacy and safety of Zestra for Women in women with and without female sexual arousal disorder. *Journal of Sex and Marital Therapy* 29(suppl. 1):33–44.

Ferguson, J. K. W. 1941. A study of the motility of the intact uterus at term. *Surgery Gynecology and Obstetrics* 73:359–366.

Ferguson, J. M. 2001. The effects of antidepressants on sexual functioning in depressed patients: a review. *Journal of Clinical Psychiatry* 62(suppl. 3):22–34.

Ferin, M. 1983. Neuroendocrine control of ovarian function in the primate. *Journal of Reproduction and Fertility* 69:369–381.

Fernández-Guasti, A., Escalante, A. L., Ahlenius, S., Hillegaart, V., & Larsson, K. 1992. Stimulation of 5-HT1A and 5-HT1B receptors in brain regions and its effects on male rat sexual behaviour. *European Journal of Pharmacology* 210:121–129.

Filippi, S., Vignozzi, L., Vannelli, G. B., Ledda, F., Forti, G., & Maggi, M. 2003. Role of oxytocin in the ejaculatory process. *Journal of Endocrinological Investigation* 26:82–86.

Filler, W., & Drezner, N. 1944. Results of surgical castration in women over forty. *American Journal of Obstetrics and Gynecology* 47:122–124.

Fischer, S. 1973. *The Female Orgasm: Psychology, Physiology, Fantasy.* New York: Basic Books.

Fisher, C., Cohen, H. D., Schiavi, R. C., Davis, D., Furman, B., Ward, K., Edwards, A., & Cunningham, J. 1983. Patterns of female sexual arousal during sleep and waking: vaginal thermo-conductance studies. *Archives of Sexual Behavior* 12:97–122.

Fox, C. A., Ismail, S., Love, D. N., Kirkham, E. E., & Loraine, J. A. 1972. Studies on the relationship between plasma testosterone levels and human sexual activity. *Journal of Endocrinology* 52:51–58.

Fox, C. A., Wolff, H. S., & Baker, J. A. 1970. Measurement of intra-vaginal and intra-uterine pressures during human coitus by radio-telemetry. *Journal of Reproduction and Fertility* 22:243–251.

Francis, S., Rolls, E. T., Bowtell, R., McGlone, F., O'Doherty, J., Browning, A., Clare, S., & Smith, E. 1999. The representation of pleasant touch in the brain and its relationship with taste and olfactory areas. *NeuroReport* 10:453–459.

Frank, E., Anderson, C., & Rubenstein, D. 1978. Frequency of sexual dysfunction in normal couples. *New England Journal of Medicine* 299:111–115.

Frazer, A., & Hensler, J. G. 1999. Serotonin. In *Basic Neurochemistry: Molecular, Cellular and Medical Aspects,* 6th ed., ed. G. S. Siegel, B. W. Agranoff, R. W. Albers, S. K. Fisher, & M. D. Uhler. Philadelphia: Lippincott Williams & Wilkins.

Freeman, W. 1958. Prefrontal lobotomy: final report of 500 Freeman and Watts patients followed for 10 to 20 years. *Southern Medical Journal* 51:739–744.

Freeman, W. 1971. Frontal lobotomy in early schizophrenia: long follow-up in 415 cases. *British Journal of Psychiatry* 119:621–624.

Freeman, W., & Watts, J. W. 1950. Psychosurgery. In *The Treatment of Mental Disorders and Intractable Pain.* Springfield, IL: Charles C Thomas.

Fugl-Meyer, K. S., Lewis, R. W., Bosch, R., Fugl-Meyer, A. R., Laumann, E. O., Lizza, E., & Martin-Morales, A. 2004. Definitions, classification, and epidemiology of sexual dysfunction. In *Sexual Medicine,* vol. 1: *Sexual Dysfunction in Men,* ed. T. Lue, F. Giuliano, S. Khoury, F. Montorsi, & R. Rosen. Paris: Health Publications.

Furuhjelm, M., Karlgren, E., & Carlstrom, K. 1984. The effect of estrogen therapy on somatic and psychiatrical symptoms in postmenopausal women. *Acta Obstetricia et Gynecologica Scandinavica* 63:655–661.

Gandhi, N., Purandare, N., & Lock, M. 1993. Surgical castration for sex offenders: boundaries between surgery and mutilation are blurred. *British Medical Bulletin* 307:1141.

Garner, W. E., & Allen, H. A. 1989. Sexual rehabilitation and heart disease. *Journal of Rehabilitation* 55:69–73.

Gerstenberg, T. C., Levin, R. J., & Wagner, G. 1990. Erection and ejaculation in man: assessment of the electromyographic activity of the bulbocavernosus and ischiocavernosus muscles. *British Journal of Urology* 65:395–402.

Ghadirian, A. M., Annable, L., & Belanger, M. C. 1992. Lithium, benzodiazepines, and sexual function in bipolar patients. *American Journal of Psychiatry* 149:801–805.

Ghadirian, A. M., Chouinard, G., & Annable, L. 1982. Sexual dysfunction and plasma prolactin levels in neuroleptic treated schizophrenic outpatients. *Journal of Nervous and Mental Disease* 170:463–467.

Ghezzi, A. 1999. Sexuality and multiple sclerosis. *Scandinavian Journal of Sexology* 2:125–140.

Giles, G. G., Severi, G., English, D. R., McCredie, M. R. E., Borland, R., Boyle, P., & Hopper, J. 2003. Sexual factors and prostate cancer. *BJU International* 92:211–216.

Gil-Vernet, J. M., Jr., Alvarez-Vijande, R., Gil-Vernet, A., & Gil-Vernet, J. M. 1994. Ejaculation in men: a dynamic endorectal ultrasonographical study. *British Journal of Urology* 73:442–448.

Girault, J. A., & Greengard, P. 1999. Principles of signal transduction. In *Neurobiology of Mental Illness,* ed. C. D. S. Charney, E. J. Nestler & B. S. Bunney. New York: Oxford University Press.

Giuliano, F., & Julia-Guilloteau, V. 2006. Neurophysiology of female genital response. In *Women's Sexual Function and Dysfunction: Study, Diag-*

nosis, and Treatment, ed. I. Goldstein, C. M. Meston, S. R. Davis, & A. M. Traish. London: Taylor & Francis.

Giuliano, F., Rampin, O., & Allard, J. 2002. Neurophysiology and pharmacology of female genital sexual response. *Journal of Sex and Marital Therapy* 28:101–121.

Goldstat, R., Briganti, E., Tran, J., Wolfe, R., & Davis, S. 2003. Transdermal testosterone improves mood, well being and sexual function in premenopausal women. *Menopause* 10:390–398.

Goldstein, I. 2002. The urologist's role in erectile dysfunction in 2002. Paper presented at New York University School of Medicine Conference. December 7.

Goldstein, I., Siroky, M. B., Sax, D. S., & Krane, R. J. 1982. Neurological abnormalities in multiple sclerosis. *Journal of Urology* 128:541–545.

Goldstein, I., Young, J. M., Fisher, J., Bangerter, K., Segerson, T., & Taylor, T. 2003. Vardenafil, a new phosphodiesterase type 5 inhibitor, in the treatment of erectile dysfunction in men with diabetes. *Diabetes Care* 26:777–783.

Gorman, D. G., & Cummings, J. L. 1992. Hypersexuality following septal injury. *Archives of Neurology* 49:308–310.

Gottesman, N. 2005. HIV over 50. *AARP: The Magazine,* July & August, 62.

Goy, R. W., & Resko, J. A. 1972. Gonadal hormones and behavior of normal and pseudohermaphroditic nonhuman female primate. *Recent Progress in Hormone Research* 28:707–733.

Graber, B., & Kline-Graber, G. 1979. Female orgasm: role of the pubococcygeus. *Journal of Clinical Psychiatry* 30:34–39.

Gräfenberg, E. 1950. The role of urethra in female orgasm. *International Journal of Sexology* 3:145–148.

Green, A. W. 1975. Sexual activity and the postmyocardial infarction patient. *American Heart Journal* 89:246–252.

Green, J. D., Clemente, C. D., & DeGroot, J. 1957. Rhinencephalic lesions and behavior in cats: an analysis of the Kluver-Bucy syndrome with particular reference to normal and abnormal sexual behavior. *Journal of Comparative Neurology* 108:505–545.

Greenblatt, R. B. 1943. Hormonal factors in libido. *Journal of Clinical Endocrinology* 3:305–306.

Gutman, M. B., Jones, D. L., & Ciriello, J. 1988. Effect of paraventricular nucleus lesions on drinking and pressor responses to ANG II. *American Journal of Physiology* 255:R882–887.

Hackbert, L., Heiman, J. R., & Meston, C. M. 1998. The effects of DHEA on sexual arousal in premenopausal women. Paper presented at the Annual Meeting of the International Academy of Sex Research, Stony Brook, NY. June 24.

Haensel, S. M., Rowland, D. L., & Slob, A. K. 1995. Serotonergic drugs and masculine sexual behavior in laboratory rats and men. In *The Pharmacology of Sexual Function and Dysfunction,* ed. J. Bancroft. Amsterdam: Excerpta Medica.

Hagemann, J. H., Berding, G., Bergh, S., Sleep, D. J., Knapp, W. H., Jonas, U., & Stief, C. G. 2003. Effects of visual sexual stimuli and apomorphine SL on cerebral activity in men with erectile dysfunction. *European Urology* 43:412–420.

Halpern, C. R., Udry, J. R., Campbell, B., Suchindran, C., & Mason, G. A. 1994. Testosterone and religiosity as predictors of sexual attitudes and activity among adolescent males: a biosocial model. *Journal of Biosocial Science* 26:217–234.

Halpern, C. R., Udry, J. R., & Suchindran, C. 1998. Monthly measures of salivary testosterone predict sexual activity in adolescent males. *Archives of Sexual Behavior* 27:445–465.

Hamann, S., Herman, R. A., Nolan, C. L., & Wallen, K. 2004. Men and women differ in amygdala response to visual sexual stimuli. *Nature Neuroscience* 7:411–416.

Harrison, W. M., Rabkin, J. G., Ehrhardt, A. A., Stewart, J. W., McGroth, P. J., Ross, D., & Quitkin, F. M. 1986. Effects of antidepressant medication on sexual function: controlled study. *Journal of Clinical Psychopharmacology* 6:144–149.

Hartman, D., Monsma, F., & Civelli, O. 1996. Interaction of antipsychotic drugs with dopamine receptors subtypes. In *Antipsychotics,* ed. J. G. Coernansky. Berlin: Springer.

Hartman, W., & Fithian, M. 1984. *Any Man Can: The Multiple Orgasmic Technique for Every Loving Man.* New York: St. Martin's Press.

Hatzichristou, D., Rosen, R. C., Broderick, G., Clayton, A., Cuzin, B., Derogatis, L., Litwin, M., Meuleman, E., O'Leary, M., Quirk, F., Sadovsky, R., & Seftel, A. 2004. Clinical evaluation and management strategy for sexual dysfunction in men and women. *Journal of Sexual Medicine* 1:49–57.

Heath, R. G. 1964. Pleasure response of human subjects to direct stimulation of the brain: physiologic and psychodynamic considerations. In *The Role of Pleasure in Behavior,* ed. R. G. Heath. New York: Harper and Row.

Heath, R. G., & Fitzjarrell, A. T. 1984. Chemical stimulation to deep forebrain nuclei in parkinsonism and epilepsy. *International Journal of Neurology* 18:163–178.

Heeb, M. M., & Yahr, P. 2001. Anatomical and functional connections among cell groups in the gerbil brain that are activated with ejaculation. *Journal of Comparative Neurology* 439:248–258.

Heller, C. G., Farney, J. P., & Myers, G. B. 1944. Development and correlation of menopausal symptoms, vaginal smear and urinary gonadotrophin changes following castration in 27 women. *Journal of Clinical Endocrinology* 4:101–108.

Hermabessiere, J., Guy, L., & Boiteaux, J. P. 1999. Human ejaculation: physiology, surgical conservation of ejaculation (in French). *Progresse Urologie* 9:305–309.

Herndon, J. G., Jr., Caggiula, A. R., Sharp, D., Ellis, D., & Redgate, E. 1978. Selective enhancement of the lordotic component of female sexual behavior in rats following destruction of central catecholamine-containing systems. *Brain Research* 141:137–151.

Higuchi, T., Uchide, K., Honda, K., & Negoro, H. 1987. Pelvic neurectomy abolishes the fetus-expulsion reflex and induces dystocia in the rat. *Experimental Neurology* 96:443–455.

Hillegaart, V. S., Ahlenius, S., & Larsson, K. 1991. Region selective inhibition of male rat sexual behavior and motor performance by localized forebrain 5HT injections: a comparison with effects produced by 8-OH-DPAT. *Behavioural Brain Research* 42:169–180.

Hilliges, M., Falconer, C., Ekman-Ordeberg, G., & Johansson, O. 1995. Innervation of the human vaginal mucosa as revealed by PGP 9.5 immunohistochemistry. *Acta Anatomica (Basel)* 153:119–126.

Hite, S. 1976. *The Hite Report.* New York: Macmillan.

Hoff, E. C., Kell, J. F., Jr., & Carroll, M. N., Jr. 1963. Effects of cortical stimulation and lesions on cardiovascular function. *Physiological Reviews* 43:68–114.

Hollander, E., & McCarley, A. 1992. Yohimbine treatment of sexual side effects induced by serotonin reuptake blockers. *Journal of Clinical Psychiatry* 53:207–209.

Hollander, X. 1981. Presentation at Fifth World Congress of Sexology, Jerusalem.

Holstege, G., Georgiadis, J. R., Paans, A. M. J., Meiners, L. C., van der Graaf, F. H. C. E., & Reinders, A. A. T. S. 2003. Brain activation during human male ejaculation. *Journal of Neuroscience* 23:9185–9193.

Hoyle, C. H. V., Stones, R. W., Robson, T., Whitley, K., & Burnstock, G. 1996. Innervation of vasculature and microvasculature of the human vagina by NOS and neuropeptide-containing nerves. *Journal of Anatomy* 188:633–644.

Hoyt, R. F., Jr. 2006. Innervation of the vagina and vulva: neurophysiology of female genital response. In *Women's Sexual Function and Dysfunction: Study, Diagnosis, and Treatment,* ed. I. Goldstein, C. M. Meston, S. R. Davis, & A. M. Traish. London: Taylor & Francis.

Hu, Z. Y., Bourreau, E., Jung-Testas, L., Robel, P., & Baulieu, E. E., 1987. Neurosteroids: oligodendrocyte mitochondria convert cholesterol to pregnenolone. *Proceedings of the National Academy of Sciences USA* 84:8215–8219.

Hubscher, C. H., & Berkley, K. J. 1994. Responses of neurons in caudal solitary nucleus of female rats to stimulation of vagina, cervix, uterine horn and colon. *Brain Research* 21:1–8.

Hull, E. M., Muschamp, J. W., & Sato, S. 2004. Dopamine and serotonin influences on male sexual behavior. *Physiology and Behavior* 83:291–307.

Hulter, B., Berne, C., & Lundberg, P. O. 1998. Sexual function in women with insulin dependent diabetes mellitus: correlation with neurological symptoms and signs. *Scandinavian Journal of Sexology* 1:43–50.

Hulter, B., & Lundberg, P. O. 1995. Sexual function in women with advanced multiple sclerosis. *Journal of Neurology, Neurosurgery, and Psychiatry* 59:83–86.

Hyde, J. S. 2005. *Biological Basis of Human Sexuality.* Washington, DC: American Psychological Association.

Hyndman, O. R., & Wolkin, J. 1943. Anterior chordotomy. *Archives of Neurology and Psychiatry* 50:129–148.

Insel, T. R., Winslow, J. T., Williams, J. R., Hastings, N., Shapiro, L. E., & Carter, C. S. 1993. The role of neurohypophyseal peptides in the central mediation of complex social processes—evidence from comparative studies. *Regulatory Peptides* 45:127–131.

Ito, T., Kawahara, K., Das, A., & Strudwick, W. 1998. The effects of ArginMax, a natural dietary supplement for enhancement of male sexual function. *Hawaii Medical Journal* 57:741–744.

Ito, T. Y., & Kawahara, K. K. 2006. A randomized, double-blind placebo-controlled clinical study on the effects of ArginMax, a natural dietary supplement for enhancement of male sexual function. Unpublished ms.

Ito, T. Y., Trant, A. S., & Polan, M. L. 2001. A double-blind placebo-controlled study of ArginMax, a nutritional supplement for enhancement of female sexual function. *Journal of Sex and Marital Therapy* 27: 541–549.

Jacobs, B. L., & Azmitia, E. C. 1992. Structure and function of the brain serotonin system. *Physiological Reviews* 72:165–229.

Jacobsen, F. M. 1992. Fluoxetine induced sexual dysfunction and an open trial of yohimbine. *Journal of Clinical Psychiatry* 53:119–122.

Jacoby, S. 1999. Great sex: What's age got to do with it? *AARP/Modern Maturity*. www.aarp.org/mmaturity/sept_oct99/greatsex.html.

Jacoby, S. 2005. Sex in America. *AARP: The Magazine,* July–August, 57–62, 114.

Janszky, J., Ebner, A., Szupera, Z., Schulz, R., Hollo, A., Szucs, A., & Clemens, B. 2004. Orgasmic aura—a report of seven cases. *Seizure* 13:441–444.

Janszky, J., Szucs, A., Halasz, P., Borbely, C., Hollo, A., Barsi, P., & Mirnics, Z. 2002. Orgasmic aura originates from the right hemisphere. *Neurology* 58:302–304.

Jarolim, L. 2000. Surgical conversion of genitalia in transsexual patients. *BJU International* 85:851–856.

Jensen, S. B. 1981. Diabetic sexual dysfunction: a comparative study of 160 insulin treated diabetic men and women in an age-matched control group. *Archives of Sexual Behavior* 10:493–504.

Jentsch, J. D., & Roth, R. H. 2000. Effects of antipsychotic drugs on dopamine release and metabolism in the central nervous system. In *Neurotransmitter Receptors in Actions of Antipsychotics Medication,* vol. 3, ed. M. Lidow. Boca Raton, FL: CRC Press.

Johnson, S. D., Phelps, D. L., & Cottler, L. B. 2004. The association of sexual dysfunction and substance use among a community epidemiological sample. *Archives of Sexual Behavior* 33:55–63.

Jones, A. K. P., Brown, W. D., Friston, K. J., Qi, L. Y., & Frackowiak, R. S. J. 1991. Cortical and subcortical localization of response to pain in man using positron emission tomography. *Proceedings of the Royal Society of London Series B* 244:39–44.

Jones, K. P., Kingsberg, S., & Whipple, B. 2005. *ARHP Clinical Proceedings: Women's Sexual Health in Midlife and Beyond.* Washington, DC: Association of Reproductive Health Professionals.

Kall, K. L. 1992. Effects of amphetamine on sexual behavior of male i.v. users in Stockholm: a pilot study. *AIDS Education and Prevention* 4:6–17.

Kaplan, H. S. 1974. *The New Sex Therapy.* New York: Quadrangle.

Kaplan, H. S. 1979. *Disorders of Sexual Drive and Other New Concepts and Techniques in Sex Therapy*. Levittown, PA: Brunner/Mazel.

Kapur, S. 2004. How antipsychotics become anti "psychotic" from dopamine to salience to psychosis. *Trends in Pharmacological Sciences* 25: 402–406.

Kapur, S., & Mamo, D. 2003. Half a century of antipsychotics and still a central role for dopamine D2 receptors. *Progress in Neuropsychopharmacology and Biological Psychiatry* 27:1081–1090.

Karama, S., Lecours, A. R., Leroux, J.-M., Bourgouin, P., Beaudoin, G., Joubert, S., & Beauregard, M. 2002. Areas of brain activation in males and females during viewing of erotic film excerpts. *Human Brain Mapping* 16:1–13.

Kassirer, J. P. 2004. *On the Take: How Medicine's Complicity with Big Business Can Endanger Your Health*. New York: Oxford University Press.

Kayner, C. E., & Sager, J. A. 1983. Breast feeding and sexual response. *Journal of Family Practice* 17:69–73.

Kenakin, T. P., Bond, R. A., & Bonner, T. I. 1992. Definition of pharmacological receptors. *Pharmacological Research* 44:351–378.

Kendrick, K. M., & Dixson, A. F. 1983. The effect of the ovarian cycle on the sexual behavior of the common marmoset (*Callithrif jacchus*). *Physiology and Behavior* 30:735–742.

Kettl, P., Zarefoss, S., Jacoby, K., Garman, C., Hulse, C., Rowley, F., Corey, R., Sredy, M., Bixler, E., & Tyson, K. 1991. Female sexuality after spinal cord injury. *Sexuality and Disability* 9:287–295.

Kim, H. L., Strelzer, J., & Gaebert, D. 1999. St. John's wort for depression: a meta-analysis of well defined clinical trials. *Journal of Nervous and Mental Disease* 187:532–538.

Kim, S. K., Park, J. H., Lee, K. C., Park, J. M., Kim, J. T., & Kim, M. C. 2003. Long-term results in patients after rectosigmoid vaginoplasty. *Plastic and Reconstructive Surgery* 112:143–151.

Kingsberg, S. A. 2002. The impact of aging on sexual function in women and their partners. *Archives of Sexual Behavior* 31:431–437.

Kingsberg, S. A., & Whipple, B. 2005. Desire: understanding female sexual response, *Health and Sexuality* 10:1–16.

Kinsey, A. C., Pomeroy, W. B., & Martin, C. E. 1948. *Sexual Behavior in the Human Male*. Philadelphia: W. B. Saunders.

Kinsey, A., Pomeroy, W., Martin, C., & Gebhard, P. 1953. *Sexual Behavior in the Human Female*. Philadelphia: W. B. Saunders.

Kirby, M., Jackson, G., Betteridge, J., & Friedli, K. 2001. Is erectile dysfunction a marker for cardiovascular disease? *International Journal of Clinical Practice* 55:614–618.

Kirchner, A., Birklein, F., Stefan, H., & Handwerker, H. 2001. Vagus nerve stimulation—a new option for the treatment of chronic pain syndromes? *Schmerz* 15:272–277.

Knowlton, L. 2000. Sexuality and aging. *Psychiatric Times* 17(1). www.psychiatrictimes.com/p000159.html.

Koeman, M., van Driel, M. F., Weijmar Schultz, W. C. M., & Mensink, H. J. A. 1996. Orgasm after radical prostatectomy. *British Journal of Urology* 77:861–864.

Koller, W. C., Vetere-Overfield, B., Williamson, A., Busenbark, K., Nash, J., & Parrish, D. 1990. Sexual dysfunction in Parkinson's disease. *Clinical Neuropharmacology* 13:461–463.

Kolodny, R. C. 1971. Sexual dysfunction in diabetic females. *Diabetes* 20:557–559.

Komisaruk, B. R. 1971. Induction of lordosis in ovariectomized rats by stimulation of the vaginal cervix: hormonal and neural interrelationships. In *Steroid Hormones and Brain Function,* ed. C. H. Sawyer & R. A. Gorski. Berkeley: University of California Press.

Komisaruk, B. R., Adler, N. T., & Hutchison, J. 1972. Genital sensory field: enlargement by estrogen treatment in female rats. *Science* 178:1295–1298.

Komisaruk, B. R., Bianca, R., Sansone, G., Gomez, L. E., Cueva-Rolon, R., Beyer, C., & Whipple, B. 1996. Brain-mediated responses to vaginocervical stimulation in spinal cord-transected rats: role of the vagus nerves. *Brain Research* 708:128–134.

Komisaruk, B. R., & Diakow, C. 1973. Lordosis reflex intensity in rats in relation to the estrous cycle, ovariectomy, estrogen administration, and mating behavior. *Endocrinology* 93:32–41.

Komisaruk, B. R., Gerdes, C., & Whipple, B. 1997. "Complete" spinal cord injury does not block perceptual responses to genital self-stimulation in women. *Archives of Neurology* 54:1513–1520.

Komisaruk, B. R., & Larsson, K. 1971. Suppression of a spinal and a cranial nerve reflex by vaginal or rectal probing in rats. *Brain Research* 35:231–235.

Komisaruk, B. R., Mosier, K. M., Criminale, C., Liu, W.-C., Zaborszky, L., Whipple, B., & Kalnin, A. J. 2002a. Functional localization of brainstem and cervical spinal cord nuclei in humans with fMRI. *American Journal of Neuroradiology* 23:609–617.

Komisaruk, B. R., & Sansone, G. 2003. Neural pathways mediating vaginal function: the vagus nerves and spinal cord oxytocin. *Scandinavian Journal of Psychology* 44:241–250.

Komisaruk, B. R., & Whipple, B. 1984. Evidence that vaginal self-stimulation in women suppresses experimentally-induced finger pain. *Society for Neuroscience Abstracts* 10:675

Komisaruk, B. R., & Whipple, B. 1991. Physiological and perceptual correlates of orgasm produced by genital or non-genital stimulation. In *Proceedings of the First International Conference on Orgasm,* ed. P. Kothari & R. Patel, Bombay: VRP Publishers.

Komisaruk, B. R., & Whipple, B. 1994. Complete spinal cord injury does not block perceptual responses to vaginal or cervical self-stimulation in women. *Society for Neuroscience Abstracts* 20:961.

Komisaruk, B. R., & Whipple, B. 1995. The suppression of pain by genital stimulation in females. *Annual Review of Sex Research* 6:151–186.

Komisaruk, B. R., & Whipple, B. 1998. Love as sensory stimulation: physiological effects of its deprivation and expression. *Psychoneuroendocrinology* 23:927–944.

Komisaruk, B. R., & Whipple, B. 2000. How does vaginal stimulation produce pleasure, pain and analgesia? In *Sex, Gender and Pain,* ed. R. B. Fillingim. Seattle: IASP Press.

Komisaruk, B. R., & Whipple, B. 2005. Brain activity imaging during sexual response in women with spinal cord injury. In *Biological Substrates of Human Sexuality,* ed. J. S. Hyde. Washington, DC: American Psychological Association.

Komisaruk, B. R., & Whipple, B. 2005. Functional MRI of the brain during orgasm in women. *Annual Review of Sex Research* 16:62–86.

Komisaruk, B. R., Whipple, B., Crawford, A., Grimes, S., Kalnin, A. J., Mosier, K., Liu, W.-C., & Harkness, B. 2002. Brain activity (fMRI and PET) during orgasm in women, in response to vaginocervical self-stimulation. Program No. 841.17 Abstract Viewer/Itinerary Planner. Washington, DC: Society of Neuroscience. CD-ROM.

Komisaruk, B., Whipple, B., Crawford, A., Grimes, S., Liu, W.-C., Kalnin, A., & Mosier, K. 2004. Brain activation during vaginocervical self-stimulation and orgasm in women with complete spinal cord injury: fMRI evidence of mediation by the vagus nerves. *Brain Research* 1024:77–88.

Komisaruk, B. R., Whipple, B., Gerdes, C., Harkness, B., & Keyes, J. W., Jr. 1997. Brainstem response to cervical self-stimulation: preliminary PET-scan analysis. *Society for Neuroscience Abstracts* 23:1001.

Kothari, P. 1989. *Orgasm: New Dimensions*. Bombay: VRP Publishers.

Kothari, P., & Patel, R., eds. 1991. *The First International Conference on Orgasm, Proceedings*. Bombay: VRP Publishers.

Kotin, J., Wilbert, D. E., Verburg, D., & Soldinger, S. M. 1976. Thioridazine and sexual dysfunction. *American Journal of Psychiatry* 133:82–85.

Kow, L., & Pfaff, D. 1973–74. Effects of estrogen treatment on the size of receptive field and response threshold of pudendal nerve in the female rat. *Neuroendocrinology* 13:299–313.

Kraemer, H. C., Becker, H. B., Brodie, H. K. H., Doering, C. H., Moas, R. H., & Hamburg, D. A. 1976. Orgasmic frequency and plasma testosterone levels in normal human males. *Archives of Sexual Behavior* 5:125–132.

Krantz, K. E. 1958. Innervation of the human vulva and vagina: a microscopic study. *Obstetrics and Gynecology* 12:382–396.

Krege, S., Bex, A., Lummen, G., & Rubben, H. 2001. Male-to-female transsexualism: a technique, results and long-term follow-up in 66 patients. *BJU International* 88:396–402.

Krüger, T. H., Haake, P., Chereath, D., Knapp, W., Jaussen, O. E., Exton, M. S., Schedlowski, M., & Hartmann, U. 2003. Specificity of the neuroendocrine response to orgasm during sexual arousal in men. *Journal of Endocrinology* 177:57–64.

Krüger, T. H. C., Haake, P., Hartmann, U., Schedlowski, M., & Exton, M. S. 2002. Orgasm-induced prolactin secretion: feedback control of sexual drive? *Neuroscience and Biobehavioral Reviews* 26:31–44.

Krüger, T. H., Hartmann, U., & Schedlowski, M. 2005. Prolactinergic and dopaminergic mechanisms underlying sexual arousal and orgasm in humans. *World Journal of Urology* 23:130–138.

Kunz, G., Beil, D., Deininger, H., Wildt, L., & Leyendecker, G. 1996. The dynamics of rapid sperm transport through the female genital tract: evidence from vaginal sonography of uterine peristalsis and hysterosalpingoscintigraphy. *Human Reproduction* 11:627–632.

Labbate, L. A., Croft, H. A., & Oleshansky, M. A. 2003. Antidepressant-related erectile dysfunction: management via avoidance, switching antidepressants, antidotes and adaptation. *Journal of Clinical Psychiatry* 64(suppl. 10):11–19.

Laborit, H., & Huguenard, P. 1951. L'hibernation artificielle par moyen pharmacodynamiques et physiques. *Presse Médicale* 59:1329.

Ladas, A. K., Whipple, B., & Perry, J. D. 1982. *The G Spot and Other Recent Discoveries about Human Sexuality.* New York: Holt, Rinehart and Winston.

Ladas, A. K., Whipple, B., & Perry, J. D. 2005. *The G Spot and Other Discoveries about Human Sexuality.* New York: Owl Books.

Landen, M., Eriksson, E., Agren, H., & Fahlen, T. 1999. Effect of buspirone on sexual dysfunction in depressed patients treated with selective serotonin reuptake inhibitors. *Journal of Clinical Psychopharmacology* 19:268–271.

Lane, R. M. 1997. A critical review of selective serotonin reuptake-inhibitor related sexual dysfunction: incidence, possible aetiology and implications for management. *Journal of Psychopharmacology (Oxford, England)* 11:72–82.

Larsson, K., & Ahlenuis, S. 1999. Brain and sexual behavior. *Annals of the New York Academy of Sciences* 877:292–308.

Larsson, K., Sodersten, P., & Beyer, C. 1973. Sexual behavior in male rats treated with estrogen in combination with dihydrotestosterone. *Hormones and Behavior* 4:289–299.

Lauerma, H. 1995. A case of moclobemide induced hyperorgasmia. *International Clinical Psychopharmacology* 10:123–124.

Laumann, E. O., Paik, A., & Rosen, R. C. 1999. Sexual dysfunction in the United States: prevalence and predictors. *JAMA: Journal of the American Medical Association* 281:537–544.

Laumann, E. O., Gagon, J. H., Michael, R. T., & Michael, S. 1994. *The Social Organization of Sexuality: Sexual Practices in the United States.* Chicago: University of Chicago Press.

Leiblum, S., Bachmann, G., Kammann, E., Calburn, D., & Schwartzman, L. 1983. Vaginal atrophy in the post-menopausal woman: the importance of sexual activity and hormones. *JAMA: Journal of the American Medical Association* 249:2195–2198.

Leitzmann, M. F., Platz, E. A., Stampfer, M. J., Willett, W. C., & Giovannucci, E. 2004. Ejaculation frequency and subsequent risk of prostate cancer. *JAMA: Journal of the American Medical Association* 291:1578–1586.

Le Vay, S. 1991. A difference in hypothalamic structure between heterosexual and homosexual men. *Science* 253:1034–1037.

Le Vay, S. 1993. *The Sexual Brain.* Cambridge, MA: MIT Press.

Levin, R. J. 1998. Sex and the human female reproductive tract—what really happens during and after coitus. *International Journal of Impotence Research* 10(suppl. 1):S14–S21.

Levin, R. J. 2002. The physiology of sexual arousal in the human female: a recreational and procreational synthesis. *Archives of Sexual Behavior* 31:405–411.

Levin, R. J. 2005. The mechanisms of human ejaculation—a critical analysis. *Sexual and Relationship Therapy* 20:123–131.

Levine, S. B. 1998. *Sexuality in Mid-Life*. New York: Plenum Press.

Levy, A. 2002. Male sexual dysfunction and the primary care physician. Paper presented at New York University School of Medicine Conference. December 7.

Lewis, R. W., Fugl-Meyer, K. S., Bosch, R., Fugl-Meyer, A. R., Laumann, E. O., Lizza, E., & Martin-Morales, A. 2004. Epidemiology/risk factors of sexual dysfunction. *Journal of Sexual Medicine* 1:35–39.

Libet, B. 1999. Consciousness: neural basis of conscious experience. In *Encyclopedia of Neuroscience,* ed. G. Adelman & B. H. Smith. New York: Elsevier.

Lief, H. I. 1977. Inhibited sexual desire. *Medical Aspects of Human Sexuality* 7:94–95.

Linde, K., Ramirez, G., Mulrow, C. D., Pauls, A., Weidenhammer, W., & Melchant, D. 1996. St. John's wort for depression an overview and meta-analysis of randomised clinical trials. *BMJ (Clinical research ed.)* 313:253–258.

Lindvall, O., Bjorklund, A., & Skagerberg, G. 1984. Selective histochemical demonstration of dopamine terminal systems in rat and telencephalon: new evidence for dopaminergic innervation of hypothalamic neurosecretory nuclei. *Brain Research* 306:19–30.

Lloyd, E. A. 2005. *The Case of the Female Orgasm: Bias in the Science of Evolution.* Cambridge, MA: Harvard University Press.

Lofving, B. 1961. Cardiovascular adjustments induced from the rostral cingulate gyrus with special reference to sympatho-inhibitory mechanisms. *Acta Physiologica Scandinavica* 53(suppl. 184):1–82.

Longcope, C., Jaffee, W., & Griffing, G. 1981. Production rates of androgens and oestrogens in post-menopausal women. *Maturitas* 3:215–223.

Lowenstein, L., Yarnitsky, D., Gruenwald, I., Deutsch, M., Sprecher, E., Gadalia, U., & Vardi, Y. 2005. Does hysterectomy affect genital sensation? *European Journal of Obstetrics, Gynecology, and Reproductive Biology* 119:242–245.

Loy, J. 1971. Estrous behavior of free ranging rhesus monkeys (*Macaca mulatta*). *Primates* 12:1–31.

Lue, T. F. 2000. Erectile dysfunction. *New England Journal of Medicine* 342:1802–1813.

Lue, T. F. 2001. Neurogenic erectile dysfunction. *Clinical Autonomic Research* 11:285–294.

Lue, T. F., Giuliano, F., Montorsi, F., Rosen, R., Andersson, K. E., Althof, S., Christ, G., Hatzichristou, D., Hirsch, M., Kimoto, Y., Lewis, R., Mc-

Kenna, K., McMahon, C., Morales, A., Mucahy, J., Padma-Nathan, H., Pryor, J., Saenz de Tejada, I., Shabsigh, R., & Wagner, G. 2004. Summary of the recommendations on sexual dysfunction in men. *Journal of Sexual Medicine* 1:6–23.

Lundberg, P. O. 1981. Sexual dysfunction in female patients with multiple sclerosis. *International Rehabilitation Medicine* 3:32–34.

Lundberg, P. O. 2005. Personal communication. August 24.

Maas, C. P., Weijengorg, P. T. M., & ter Kuile, M. M. 2003. The effect of hysterectomy on sexual functioning. *Annual Review of Sex Research* 14:83–113.

Macfarlane, I., Bliss, M., Jackson, J. G. L., & Williams, G. 1997. The history of diabetes. In *Textbook of Diabetes*, ed. J. Pickup & G. Williams. Oxford: Blackwell Science.

MacLean, P. D. 1952. Some psychiatric implications of physiological studies on frontotemporal portion of limbic system (visceral brain). *Electroencephalography and Clinical Neurophysiology Supplement* 4:407–418.

MacLean, P. D. 1955. The limbic system ("visceral brain") and emotional behavior. *American Medical Association Archives of Neurology and Psychiatry* 73:130–134.

Mah, K., & Binik, Y. M. 2001. The nature of human orgasm: a critical review of major trends. *Clinical Psychology Review* 21:823–856.

Mah, K., & Binik, Y. M. 2005. Are orgasms in the mind or the body? Psychosocial versus physiological correlates of orgasmic pleasure and satisfaction. *Journal of Sex and Marital Therapy* 31:187–200.

Maines, R. 1989. Socially camouflaged technologies: the case of the electromechanical vibrator. *Technology and Society Magazine, IEEE* 8:3–11, 23.

Maixner, W., & Randich, A. 1984. Role of the right vagal nerve trunk in antinociception. *Brain Research* 298:374–377.

Malatesta, V. J., Pollack, R. H., Crotty, T. D., & Peacock, L. J. 1982. Acute alcohol intoxication and female orgasmic response. *Journal of Sex Research* 18:1–17.

Malmnas, W. 1973. Monoaminergic influence on testosterone-activating copulatory behavior in the castrated male rats. *Acta Physiologica Scandinavica Supplementum* 395:1–118.

Mantzaros, C., Georgiadis, E. J., & Trichopoulas, D. 1995. Contribution of dihydrotestosterone to male sexual behavior. *BMJ (Clinical research ed.)* 310:1289–1291.

Maravilla, K. R. 2006. Blood flow: magnetic resonance imaging and brain imaging for evaluating sexual arousal in women. In *Women's Sexual Func-*

tion and Dysfunction: Study, Diagnosis, and Treatment, ed. I. Goldstein, C. M. Meston, S. R. Davis, & A. M. Traish. London: Taylor & Francis.

Maravilla, K. R., Cao, Y., Heiman, J. R., Yang, C., Garland, P. A., Peterson, B. T., & Carter, W. O. 2005. Noncontrast dynamic magnetic resonance imaging for quantitative assessment of female sexual arousal. *Journal of Urology* 173:162–166.

Marberger, H. 1974. The mechanisms of ejaculation. In *Physiology and Genetics of Reproduction,* ed. E. M. Coutinho & F. Fuchs. New York: Plenum Press.

Margolis, J. 2004. *O: The Intimate History of the Orgasm.* New York: Grove Press.

Marks, L. S., Duda, C., Dorey, F. J., Macairan, M. L., & Santos, P. B. 1999. Treatment of erectile dysfunction with sildenafil. *Urology* 53:19–24.

Marshall, J. F., Turner, B. H., & Teitelbaum, P. 1971. Sensory neglect produced by lateral hypothalamic damage. *Science* 174:523–525.

Martinez-Gomez, M., Chirino, R., Beyer, C., Komisaruk, B. R., & Pacheco, P. 1992. Visceral and postural reflexes evoked by genital stimulation in urethane-anesthetized female rats. *Brain Research* 575:279–284.

Mas, M., González-Mora, J. L., Louilot, A., Solé, C., & Guadalupe, T. 1990. Increased dopamine release in the nucleus accumbens of copulating male rats as evidenced by in vivo voltammetry. *Neuroscience Letters* 110:303–308.

Masters, W., & Johnson, V. 1966. *Human Sexual Response.* Boston: Little, Brown.

Masters, W., & Johnson, V. 1970. *Human Sexual Inadequacy.* Boston: Little Brown.

Matsuhashi, M., Maki, A., Takanami, M., Fujio, K., Miura, K., Nakayoma, K., Shirai, M., & Ando, K. 1984. Clinical experience of bromazepam for psychogenic impotence patients. *Hinyokika Kiyo: Acta Urologica Japonica* 30:1697–1701.

Maurice, W. L. 1999. *Sexual Medicine in Primary Care.* St. Louis, MO: Mosby.

Mazer, N. A. 2000. Transdermal testosterone treatment in women with impaired sexual function after oöphorectomy. *New England Journal of Medicine* 343:682–688.

McCabe, M. P. 2002. Relationship functioning and sexuality among people with multiple sclerosis. *Journal of Sex Research* 39:302–309.

McCann, S. M., & Ojeda, S. R. 1996. The anterior pituitary and hypothalamus. In *Textbook of Endocrine Physiology,* ed. J. E. Griffin & S. R. Ojeda. Oxford: Oxford University Press.

McCoy, N., & Davidson, J. M. 1985. A longitudinal study of the effects of menopause on sexuality. *Maturitas* 7:203–210.

McDonald, P. C. 1971. Dynamics of androgen and estrogen secretion. In *Control of Gonadal Steroid Secretion,* ed. D. T. Baird & J. A. Strong. Edinburgh: Edinburgh University Press.

McKenna, K. E. 2005. The central control and pharmacological modulation of sexual function. In *Biological Substrates of Human Sexuality,* ed. J. S. Hyde. Washington, DC: American Psychological Association.

McKenzie, K. G., & Proctor, L. D. 1946. Bilateral frontal lobe leucotomy in the treatment of mental disease. *Canadian Medical Association Journal* 55:433–439.

McMahon, C. G. 1998. Treatment of premature ejaculation with sertraline hydrochloride: single blind placebo controlled cross over study. *Journal of Urology* 159:1935–1938.

McMahon, C. G., Abdo, C., Incrocci, L., Perelman, M., Rowland, D., Waldinger, M., & Zhong, C. X. 2004. Disorders of orgasm and ejaculation in men. *Journal of Sexual Medicine* 1:58–65.

McMahon, C. G., & Touma, K. 1999a. Treatment of premature ejaculation with paroxetine hydrochloride. *International Journal of Impotence Research* 11:241–245.

McMahon, C. G., & Touma, K. 1999b. Treatment of premature ejaculation with paroxetine hydrochloride as needed: 2 single blind placebo controlled cross over studies. *Journal of Urology* 161:1826–1830.

McPherson, K., Herbert, A., Judge, A., Clarke, A., Bridgman, S., Maresh, M., & Overton, C. 2005. Psychosexual health 5 years after hysterectomy: population-based comparison with endometrial ablation for dysfunctional uterine bleeding. *Health Expectations* 8:234–243.

Meaddough, E. L., Olive, D. L., Gallup, P., Perlin, M., & Kliman, H. J. 2002. Sexual activity, orgasm and tampon use are associated with a decreased risk for endometriosis. *Gynecologic and Obstetric Investigation* 53:163–169.

Melis, M. R., & Argiolas, A. 1993. Nitric oxide synthase inhibitors prevent apomorphine and oxytocin-induced penile erection and yawning in male rats. *Brain Research Bulletin* 32:71–74.

Melis, M. R., & Argiolas, A. 1995. Dopamine and sexual behavior. *Neuroscience and Biobehavioral Reviews* 19:19–38.

Melis, M. R., Mauri, A., & Argiolas, A. 1994. Apomorphine and oxytocin induced penile erection and yawning in intact and castrated male rats: effect of sexual steroids. *Neuroendocrinology* 59:349–354.

Melis, M. R., Succu, S., Mascia, M. S., & Argiolas, A. 2005. PD-168077, a selective dopamine D4 receptor agonist, induces penile erection when injected into the paraventricular nucleus of male rats. *Neuroscience Letters* 379:59–62.

Melkersson, K. 2005. Differences in prolactin elevation and related symptoms of atypical antipsychotics in schizophrenic patients. *Journal of Clinical Psychiatry* 66:761–767.

Meloy, S. 2006. ABC News. http://abcnews.go.com/GMA/Living/story?id=235788&page=1.

Mendelson, C. R. 1996. Mechanisms of hormone action. In *Textbook of Endocrine Physiology*, ed. J. E. Griffin & S. R. Ojeda. Oxford: Oxford University Press.

Messe, M. R., & Geer, J. H. 1985. Voluntary vaginal musculature contractions as an enhancer of sexual arousal. *Archives of Sexual Behavior* 14:13–28.

Meston, C. M., & Frohlich, P. F. 2000. The neurobiology of sexual function. *Archives of General Psychiatry* 57:1012–1030.

Meston, C. M., & Heiman, J. R. 2002. Acute dehydroepiandrosterone effects on sexual arousal in premenopausal women. *Journal of Sex and Marital Therapy* 28:53–60.

Meston, C. M., Hull, E., Levin, R. J., & Sipski, M. 2004. Disorders of orgasm in women. *Journal of Sexual Medicine* 1:66–68.

Meston, C. M., Levin, R., Sipski, M. L., Hull, E. M., & Heiman, J. R. 2004. Women's orgasm. *Annual Review of Sex Research* 25:173–257.

Meuwissen, I., & Over, R. 1992. Sexual arousal across phases of the human menstrual cycle. *Archives of Sexual Behavior* 21:101–119.

Michael, R. P., & Welegalla, J. 1968. Ovarian hormones and the sexual behaviour of the female rhesus monkey (*Macaca mulatta*) under laboratory conditions. *Journal of Endocrinology* 41:407–420.

Michael, R. P., & Wilson, M. 1973. Effects of castration and hormone replacement in fully adult male rhesus monkeys (*Macaca mulatta*). *Endocrinology* 95:150–159.

Michael, R. P., Zumpe, D., & Bonsall, R. W. 1986. Comparison of the effects of testosterone and dihydrotestosterone on the behavior of male cynomolgus monkeys (*Macaca fascicularis*). *Physiology and Behavior* 36:349–355.

Miller, B. L., Cummings, J. L., McIntyre, H., Ebers, G., & Grode, M. 1986. Hypersexuality or altered sexual preference following brain injury. *Journal of Neurology, Neurosurgery, and Psychiatry* 49:867–873.

Miller, N. E. 1938. Old minds rejuvenated by sex hormones. *Science News Letter* 24:201.

Miller, N. S., & Gold, M. S. 1988. The human sexual response and alcohol and drugs. *Journal of Substance Abuse Treatment* 5:171–177.

Minderhoud, J. M., Leemhuis, J. G., Kremer, J., Laban, E., & Smits, P. M. L. 1984. Sexual disturbances arising from multiple sclerosis. *Acta Neurologica Scandinavica* 70:299–306.

Mirin, S. M., Meyer, R. F., Mendelson, J. H., & Ellingboe, J. 1980. Opiate use and sexual function. *American Journal of Psychiatry* 137:909–915.

Mitchell, J. E., & Popkin, M. K. 1983. The pathophysiology of sexual dysfunction associated with antipsychotic drug therapy in males: a review. *Archives of Sexual Behavior* 12:173–183.

Money, J. 1960. Phantom orgasm in the dreams of paraplegic men and women. *Archives of General Psychiatry* 3:373–382.

Money, J. 1970. Use of androgen depleting hormone in the treatment of male sex offenders. *Journal of Sex Research* 6:165–172.

Money, J., Wainwright, G., & Hingburger, D. 1991. *The Breathless Orgasm: A Lovemap Biography of Asphyxiophilia.* New York: Prometheus Books.

Monnier, M. 1968. *Functions of the Nervous System.* New York: Elsevier.

Montejo, A. L., Llorca, G., Izquierdo, J. A., & Rico-Villademoros, F. 2001. Incidence of sexual dysfunction associated with antidepressant agents: a prospective multicenter study of 1022 outpatients. *Journal of Clinical Psychiatry* 62(suppl. 3):10–21.

Moody, K. M., Steinman, J. L., Komisaruk, B. R., & Adler, N. T. 1994. Pelvic neurectomy blocks oxytocin-facilitated sexual receptivity in rats. *Physiology and Behavior* 56:1057–1060.

Morales, A., Condra, M., Owen, J., Surridge, D., Fenemore, J., & Harris, C. 1987. Is yohimbine effective in the treatment of organic impotence? Results of a controlled trial. *Journal of Urology* 137:1168–1172.

Moralí, G., Larsson, K., & Beyer, C. 1977. Inhibition of testosterone induced sexual behavior in the castrated male rat by aromatase blockers. *Hormones and Behavior* 9:203–213.

Mos, J., Mollet, I., Tolboom, J. T. B. M., Waldinger, M. D., & Olivier, B. 1999. A comparison of the effects of different serotonin reuptake blockers on sexual behavior of the male rat. *European Neuropsychopharmacology* 9:123–135.

Mould, D. E. 1980. Neuromuscular aspects of women's orgasms. *Journal of Sex Research* 16:193–201.

Mulhall, J. P., & Goldstein, I. 1996. Epidemiology of erectile dysfunction. In *Diagnosis and Management of Male Sexual Dysfunction,* ed. J. J. Mulcahy. New York: Igaku-Shoin.

Muller, W. E., Singer, A., Wonnemann, M., Hafner, U., Rolli, M., & Schafer, C. 1998. Hyperforin represents the neurotransmitter reuptake inhibiting constituent of hypericum extract. *Pharmacopsychiatry* 31(suppl. 1):81–85.

Murphy, M. R., Checkley, S. A., Seckl, J. R., & Lightman, S. L. 1990. Naloxone inhibits oxytocin release at orgasm in man. *Journal of Clinical Endocrinology and Metabolism* 71:1056–1058.

Murrell, T. G. 1995. The potential for oxytocin (OT) to prevent breast cancer: a hypothesis. *Breast Cancer Research and Treatment* 35:225–229.

Nadler, R. D. 1975. Sexual cyclicity in captive lowland gorillas. *Science* 189:813–814.

Naftolin, F., Ryan, K. J., Davies, L. J., Reddy, V. V., Flores, F., Petro, L., Kuhn, M., White, R. J., Takaoka, Y., & Wolin, L. 1975. The transformation of estrogens by central neuroendocrine tissues. *Recent Progress in Hormone Research* 31:295–319.

Nagle, C. A., & Denari, J. H. 1983. The cebus monkey (*Cebus apella*). In *New World Primates,* ed. J. P. Hearn. Lancaster, UK: MTP Press.

Nathorst-Boos, J., von Schoultz, B., & Carlstrom, K. 1993. Effective ovarian removal and estrogen replacement therapy: effects on sexual life, psychological well being, and androgen status. *Journal of Psychosomatic Obstetrics and Gynaecology* 14:283–293.

Nauta, W. J. H. 1972. Neural associations of the frontal cortex. *Acta Neurobiologiae Experimentalis* 32:125–140.

Nauta, W. J. H., & Feirtag, M. 1986. *Fundamental Neuroanatomy.* New York: W. H. Freeman.

Ness, T. J., Randich, A., Fillingim, R., Faught, R. E., & Backensto, E. M. 2001. Left vagus nerve stimulation suppresses experimentally induced pain. *Neurology* 56:985–986.

Nestler, J. E., & Duman, R. S. 1998. G proteins. In *Basic Neurochemistry,* ed. G. J. Siegel, B. W. Agranoff, R. Wayne Albers, S. K. Fisher, & M. D. Anduhler. Baltimore: Lippincott Williams and Wilkins.

Netter, F. H. 1986. *The Ciba Collection of Medical Illustrations. Nervous System. Part I: Anatomy and Physiology.* Summit, NJ: Ciba Pharmaceutical.

Neubig, R. R., & Thomsen, W. J. 1989. How does a key fit a flexible lock? Structure and dynamics in receptor function. *Bioessays* 11:136–141.

Newton, N. 1955. *Maternal Emotions: A Study of Women's Feelings toward Menstruation, Pregnancy, Childbirth, Breast Feeding, Infant Care and Other Aspects of Their Femininity.* New York: Paul B. Hoeber.

Nieschlag, E. 1979. The endocrine function of the human testis in regard to sexuality. In *Sex, Hormone and Behaviour,* Ciba Foundation Symposia, vol. 62. Amsterdam: Excerpta Medica.

Noble, J. 1996. *Textbook of Primary Care Medicine.* St. Louis, MO: Mosby.

Norden, M. J. 1994. Buspirone treatment of sexual dysfunction associated with selective serotonin re-uptake inhibitors. *Depression* 2:109–112.

O'Connor, D. B., Archer, J., & Woo, F. C. 2004. Effects of testosterone on mood, aggression, and sexual behavior in young men: a double-blind, placebo-controlled, cross-over study. *Journal of Clinical Endocrinology and Metabolism* 89:2837–2845.

Odent, M. 1999. *The Scientification of Love.* London: Free Association Books.

Ogawa, S., Lee, T. M., Kay, A. R., & Tank, D. W. 1990. Brain magnetic resonance imaging with contrast dependent on blood oxygenation. *Proceedings of the National Academy of Sciences USA* 87:9868–9872.

Ortega-Villalobos, M., Garcia-Bazan, M., Solano-Flores, L. P., Ninomiya-Alarcon, J. G., Guevara-Guzman, R., & Wayner, M. J. 1990. Vagus nerve afferent and efferent innervation of the rat uterus: an electrophysiological and HRP study. *Brain Research Bulletin* 25:365–371.

Ottesen, B., Pedersen, B., Nielsen, J., Dalgaard, D., Wagner, G., & Fahrenkrug, J. 1987. Vasoactive intestinal polypeptide (VIP) provokes vaginal lubrication in normal women. *Peptides* 8:797–800.

Paget, L. 2001. *The Big* O. New York: Broadway Books.

Papez, J. W. 1937. A proposed mechanism of emotion. *Archives of Neurology and Psychiatry* 38:725–743.

Park, K., Kang, H. K., Seo, J. J., Kim, H. J., Ryu, S. B., & Jeong, G. W. 2001. Blood-oxygenation-level-dependent functional magnetic resonance imaging for evaluating cerebral regions of female sexual arousal response. *Urology* 57:1189–1194.

Penfield, W. 1958. Functional localization in temporal and deep Sylvian areas. *Research Publications—Association for Research in Nervous and Mental Disease* 36:210–226.

Penfield, W., & Faulk, M. E., Jr. 1955. The insula: further observations on its function. *Brain* 78:445–470.

Penfield, W., & Rasmussen, T. 1950. *The Cerebral Cortex of Man.* New York: Macmillan.

Perelman, M. 2001. Sildenafil, sex therapy, and retarded ejaculation. *Journal of Sex Education and Therapy* 26:13–21.

Perot, P., & Penfield, W. 1960. Hallucinations of past experience and experiential responses to stimulation of temporal cortex. *Transactions of the American Neurological Association* 85:80–84.

Perry, J. D., & Whipple, B. 1981. Pelvic muscle strength of female ejaculators: evidence in support of a new theory of orgasm. *Journal of Sex Research* 17:22–39.

Perry, J. D., & Whipple, B. 1982. Multiple components of the female orgasm. In *Circumvaginal Musculature and Sexual Function,* ed. B. Graber. New York: S. Karger.

Persky, H., Charney, N., Lief, H. I., O'Brien, C. P., Miller, W. R., & Strauss, D. 1978. The relationship of plasma estradiol level to sexual behavior in young women. *Psychosomatic Medicine* 40:523–535.

Persky, H., Lief, H. I., Strauss, D., Miller, W. R., & O'Brien, C. P. 1978. Plasma testosterone level and sexual behavior of couples. *Archives of Sexual Behavior* 7:157–173.

Persky, H., O'Brien, C. P., & Kahn, M. A. 1976. Reproductive hormone levels, sexual activity and moods during the menstrual cycle. *Psychosomatic Medicine* 38:62–63.

Peters, L. C., Kristal, M. B., & Komisaruk, B. R. 1987. Sensory innervation of the external and internal genitalia of the female rat. *Brain Research* 408:199–204.

Petridou, E., Giokas, G., Kuper, H., Mucci, L. A., & Trichopoulos, D. 2000. Endocrine correlates of male breast cancer risk: a case-control study in Athens, Greece. *British Journal of Cancer* 83:1234–1237.

Pfaus, J. G., Damsma, G., Wenkstern, D., & Fibiger, H. C. 1995. Sexual activity increases dopamine transmission in the nucleus accumbens and striatum of female rats. *Brain Research* 693:21–30.

Pfaus, J. G., & Gorzalka, B. B. 1987. Opioids and sexual behavior. *Neuroscience and Biobehavioral Reviews* 11:1–34.

Pfaus, J. G., & Heeb, M. M. 1997. Implications of immediate-early gene induction in the brain following sexual stimulation of female and male rodents. *Brain Research Bulletin* 44:397–407.

Philipp, M., Kohnen, R., & Benkert, O. 1993. A comparison study of moclobemide and doxepin in major depression with special reference to effects on sexual dysfunction. *International Clinical Psychopharmacology* 7:149–153.

Phillips, A. G., Pfaus, J. G., & Blaha, C. D. 1991. Dopamine and motivated behavior: insights provided by in vivo analysis. In *Mesocorticolimbic Dopamine System: From Motivation to Action,* ed. P. Willner & J. Scheel-Kruger. London: Wiley.

Phillips, N. A. 2000. Female sexual dysfunction: evaluation and treatment. *American Family Physician* 62:127–136, 141–142.

Phoenix, C. H. 1973. The role of testosterone in the sexual behavior of laboratory male rhesus. In *Symposium of the IVth International Congress of Primatology,* vol. 2, ed. C. H. Phoenix. Basel: Karger.

Phoenix, C. H. 1974. Effects of dihydrotestosterone on sexual behavior of castrated male rhesus monkeys. *Physiology and Behavior* 12:1045–1055.

Pilowsky, L. S., Mulligan, R. S., Acton, P. D., Ell, P. J., Casta, D. C., & Kerwin, R. W. 1997. Limbic selectivity of clozapine. *Lancet* 350:490–491.

Plaut, S. M., Graziottin, A., & Heaton, J. P. W. 2004. *Fast Facts—Sexual Dysfunction.* Oxford: Health Press.

Pleim, E. T., Matochik, J. A., Barheld, R. J., & Auerbach, S. B. 1990. Correlation of dopamine release in the nucleus accumbens with masculine sexual behavior in rats. *Brain Research* 524:160–163.

Ploner, M., Gross, J., Timmermann, L., & Schnitzler, A. 2002. Cortical representation of first and second pain sensation in humans. *Proceedings of the National Academy of Sciences USA* 99:12444–12448.

Poline, J. B., Holmes, A., Worsley, K., & Friston, K. J. 1997. Making statistical inferences. In *Human Brain Function,* ed. R. S. J. Frackowiak, K. J. Friston, C. D. Frith, R. J. Dolan & J. C. Mazziotta. New York: Academic Press.

Preskorn, S. H. 1995. Comparison of the tolerability of bupropion, fluoxetine, imipramine, nefazodone, paroxetine, sertraline and venlafaxine. *Journal of Clinical Psychiatry* 50(suppl. 6):12–21.

Ramachandran, V. S., & Blakeslee, S. 1999. *Phantoms in the Brain: Human Nature and the Architecture of the Mind.* London: Fourth Estate.

Randich, A., & Gebhart, G. F. 1992. Vagal afferent modulation of nociception. *Brain Research* 17:77–99.

Reading, P. J., & Will, R. G. 1997. Unwelcome orgasms. *Lancet* 350:1746.

Redoute, J., Stoleru, S., Gregoire, M.-C., Costes, N., Cinotti, L., Lavenne, F., Le Bars, D., Forest, M. G., & Pujol, J.-F. 2000. Brain processing of visual sexual stimuli in human males. *Human Brain Mapping* 11:162–177.

Reiman, E. M., Lane, R. D., Ahern, G. L., Schwartz, G. E., Davidson, R. J., Friston, K. J., Yun, L. S., & Chen, K. 1997. Neuroanatomical correlates of externally and internally generated human emotion. *American Journal of Psychiatry* 154:918–925.

Reynolds, R. D. 1997. Sertraline induced anorgasmia treated with intermittent refazodone (letter). *Journal of Clinical Psychiatry* 58:89.

Riley, A. J. 1999. Life-long absence of sexual drive in a woman associated with 5α-dihydrotestosterone deficiency. *Journal of Sex and Marital Therapy* 25:73–78.

Robbins, M. B., & Jensen, G. 1978. Multiple orgasms in males. *Journal of Sex Research* 14:21–26.

Robinson, P. 1976. *The Modernization of Sex*. London: Paul Elek.

Rodgers, J. E. 2001. *Sex: A Natural History.* New York: Henry Holt.

Rodriguez-Sierra, J. F., Crowley, W. R., & Komisaruk, B. R. 1975. Vaginal stimulation induces sexual receptivity to males, and prolonged lordosis responsiveness in rats. *Journal of Comparative and Physiological Psychology* 89:79–85.

Rosen, R. C. 2000. Prevalence and risk factors of sexual dysfunction in men and women. *Current Psychiatry Reports* 2:189–195.

Rosen, R. C. 2002. Assessment of female sexual dysfunction: review of validated methods. *Fertility and Sterility* 77:S89–S93.

Rosen, R., Lane, R., & Menza, V. 1999. Effects of SSRIs on sexual function: a critical review. *Clinical Psychopharmacology* 19:67–86.

Rosenblatt, J., & Aronson, L. R. 1958. The decline of sexual behavior in male cats after castration with special reference to the role of prior sexual experience. *Behaviour* 12:285–338.

Roussis, N. P., Waltrous, L., Kerr, A., Robertazzi, R., & Cabbad, M. F. 2004. Sexual response in the patient after hysterectomy: total abdominal versus supracervical versus vaginal procedure. *American Journal of Obstetrics and Gynecology* 190:1427–1428.

Rowe, D. W., & Erskine, M. S. 1993. C-fos proto-oncogene activity induced by mating in the preoptic area, hypothalamus and amygdala in the female rat: role of afferent input via the pelvic nerve. *Brain Research* 621:25–34.

Rowland, D. L., Kallan, K. H., & Slob, A. K. 1997. Yohimbine, erectile capacity, and sexual response in men. *Archives of Sexual Behavior* 26:49–62.

Rowland, D. L., & Tai, W. 2003. A review of plant-derived and herbal approaches to the treatment of sexual dysfunction. *Journal of Sex and Marital Therapy* 29:185–205.

Safi, A. M., & Stein, R. A. 2001. Cardiovascular risks of sexual activity. *Current Psychiatry Reports* 3:209–214.

Sakakibara, R., Shinotoh, H., Uchiyama, T., Sakuma, M., Kashiwado, M.,

Yoshiyama, M., & Hattori, T. 2001. Questionnaire-based assessment of pelvic orgasm dysfunction in Parkinson's disease. *Autonomic Neuroscience: Basic and Clinical* 17:76–85.

Salmon, V. J., & Geist, S. H. 1943. The effect of androgens upon libido in women. *Journal of Clinical Endocrinology* 3:235–238.

Sansone, G., & Komisaruk, B. R. 2001. Evidence that oxytocin is an endogenous stimulator of autonomic sympathetic preganglionics: the pupillary dilatation response to vaginocervical stimulation in the rat. *Brain Research* 898:265–271.

Sapolsky, R. M. 1985. Stress induced suppression of testicular function in the wild baboon: role of glucocorticoids. *Endocrinology* 116:2273–2278.

Sarrel, P. M. 1999. Psychosexual effects of menopause: role of androgens. *American Journal of Obstetrics and Gynecology* 180:319–324.

Sayle, A. E., Savitz, D. A.,Thorp, J. M., Jr., Hertz-Picciotto, I., & Wilcox, A. J. 2001. Sexual activity during late pregnancy and risk of preterm delivery. *Obstetrics and Gynecology* 97:283–289.

Schally, A. V. 1978. Aspects of hypothalamic regulation of the pituitary gland. *Science* 202:390–402.

Schiavi, R. C., Schreiner-Engel, P., Mandeli, J., Schanzer, H., & Cohen, E. 1990. Healthy aging and male sexual function. *American Journal of Psychiatry* 147:766–771.

Schiavi, R. C., Stimmel, B. B., Mandeli, J., & Rayfield, E. J. 1993. Diabetes mellitus and male sexual function. *Diabetologia* 36:665–675.

Schiavi, R. C., & White, D. 1976. Androgen and male sexual function: a review of human studies. *Journal of Sex and Marital Therapy* 2:214–228.

Schindler, A. E. 1975. Steroid metabolism of fetal tissues. II: Conversion of androstenedione to estrone. *American Journal of Obstetrics and Gynecology* 123:265–268.

Schover, L. R., Thomas, A. J., Lakin, M. M., Montague, D. K., & Fischer, J. 1988. Orgasm phase dysfunction in multiple sclerosis. *Journal of Sex Research* 25:548–554.

Schreiner, L., & Kling, A. 1953. Behavioral changes following rhinencephalic injury in cat. *Journal of Neurophysiology* 16:643–659.

Schreiner-Engel, P. 1980. Female sexual arousability: its relation to gonadal hormones and the menstrual cycle. Ph.D. diss., New York University.

Schreiner-Engel, P. 1983. Diabetes mellitus and female sexuality. *Sexuality and Disability* 6:83–92.

Schreiner-Engel, P., Schiavi, R. C., Smith, H., & White, D. 1981. Sexual arousability and the menstrual cycle. *Psychosomatic Medicine* 43:199–214.

Schreiner-Engel, P., Schiavi, R. C., White, D., & Ghizzani, A. 1989. Low sexual desire in women: the role of reproductive hormones. *Hormones and Behavior* 23:221–234.

Schultz, W. 2000. Multiple reward signals in the brain. *Nature Reviews: Neuroscience* 1:199–207.

Schwartz, M. B., Bauman, J. E., & Masters, W. H. 1982. Hyperprolactinemia and sexual disorders in men. *Biological Psychiatry* 17:861–876.

Schwartz, R. H., Milteer, R., & Le Beau, M. A. 2000. Drug-facilitated sexual assault ("date rape"). *Southern Medical Journal* 93:558–561.

Scura, K. W., & Whipple, B. 1995. HIV infection and AIDS in the elderly. In *Gerontological Nursing*, ed. M. Stanley & P. G. Beare. Philadelphia: F. A. Davis.

Seecof, R., & Tennant, F. S., Jr. 1986. Subjective perceptions to the intravenous "rush" of heroin and cocaine in opiate addicts. *American Journal of Drug and Alcohol Abuse* 12:79–87.

Segraves, R. T. 1992. Overview of sexual dysfunction complicating the treatment of depression. *Journal of Clinical Psychiatry* 53(suppl. 10B):4–10.

Segraves, R. T. 1993. Treatment-emergent sexual dysfunction in affective disorder: a review and management strategies. *Journal of Clinical Psychiatry Monograph* 11:57–60.

Segraves, R. T. 1995. Yohimbine may alleviate sexual dysfunction. *Psychopharmacology Update* 6:4.

Segraves, R. T., Clayton, A., Croft, H., Wolf, A., & Warnock, J. 2004. Bupropion sustained release for the treatment of hypoactive sexual desire disorder in premenopausal women. *Journal of Clinical Psychopharmacology* 24:339–342.

Semans, J. 1956. Premature ejaculation: new approach. *Southern Medical Journal* 49:353–358.

Sem-Jacobsen, C. W. 1968. *Depth-Electrographic Stimulation of the Human Brain and Behavior.* Springfield, IL: Charles C Thomas.

Shafik, A. 1998. The mechanism of ejaculation: the glans-vasal and urethromuscular reflexes. *Archives of Andrology* 41:71–78.

Shafik, A. 2000. Mechanism of ejection during ejaculation: identification of a urethrocavernosus reflex. *Archives of Andrology* 44:77–83.

Shen, W. W. 1997. The metabolism of psychoactive drugs: a review of enzymatic, biotransformation and inhibition. *Biological Psychiatry* 41:814–826.

Shen, W. W., & Sata, L. S. 1990. Inhibited female orgasm resulting from psychotropic drugs: a five year updated clinical review. *Journal of Reproductive Medicine* 35:11–14.

Shen, W. W., Urosevich, A., & Clayton, D. W. 1999. Sildenafil in the treatment of female sexual dysfunction induced by selective serotonin reuptake inhibitors. *Journal of Reproductive Medicine* 44:535–542.

Sherwin, B. B. 1988. A comparative analysis of the role of androgen in human male and female sexual behavior: behavioral specificity, critical thresholds, and sensitivity. *Psychobiology* 16:416–425.

Sherwin, B. B. 1993. Sexuality and the menopause. In *The Modern Management of the Menopause,* ed. G. Berg & M. Hammer. New York: Parthenon.

Sherwin, B. B., & Gelfand, M. M. 1987. The role of androgen in the maintenance of sexual functioning in oophorectomized women. *Psychosomatic Medicine* 49:397–409.

Sherwin, B. B., Gelfand, M. M., & Brender, W. 1985. Androgen enhances sexual motivation in females: a prospective, cross-over study of sex steroid administration in the surgical menopause. *Psychosomatic Medicine* 47:339–351.

Shifren, J. L., Braunstein, G. D., Simon, J. A., Casson, P. R., Buster, J. E., Redmond, G. P., Burki, R. E., Ginsburg, E. S., Rosen, R. C., Leiblum, S. R., Caramelli, K. E., & Mazer, N. A. 2000. Transdermal testosterone treatment in women with impaired sexual function after oophorectomy. *New England Journal of Medicine* 343:682–688.

Simmons, D. A., & Yahr, P. 2002. Projections of the posterodorsal preoptic nucleus and the lateral part of the posterodorsal medial amygdala in male gerbils, with emphasis on cells activated with ejaculation. *Journal of Comparative Neurology* 444:75–94.

Simpson, J. A., & Weiner, E. S. C., eds. 2002a. Orgasm. *Oxford English Dictionary.* Oxford: Clarendon Press.

Simpson, J. A., & Weiner, E. S. C., eds. 2002b. Paraphilia. *Oxford English Dictionary.* Oxford: Clarendon Press.

Singer, I. 1973. *The Goals of Sexuality.* New York: Schocken.

Singer, J., & Singer, I. 1972. Types of female orgasm. *Journal of Sex Research* 8:255–267.

Singh, D., Meyer, W., Zambarano, R. J., & Hurlbert, D. F. 1998. Frequency and timing of coital orgasm in women desirous of becoming pregnant. *Archives of Sexual Behavior* 27:15–29.

Sipski, M. L. 2001. Sexual response in women with spinal cord injury: neurologic pathways and recommendations for the use of electrical stimulation. *Journal of Spinal Cord Medicine* 24:155–158.

Sipski, M. L., & Alexander, C. J. 1995. Spinal cord injury and female sexuality. *Annual Review of Sex Research* 6:224–244.

Sipski, M., Alexander, C., & Rosen, R. 1995. Orgasm in women with spinal cord injuries: a laboratory-based assessment. *Archives of Physical Medicine and Rehabilitation* 76:1097–1102.

Sipski, M. L., Alexander, C. J., & Rosen, R. 2001. Sexual arousal and orgasm in women: effects of spinal cord injury. *Annals of Neurology* 49: 35–44.

Sipski, M. L., Komisaruk, B., Whipple, B., & Alexander, C. J. 1993. Physiologic responses associated with orgasm in the spinal cord injured female. *Archives of Physical Medicine and Rehabilitation* 74:1270.

Sirpurapu, K. B., Gupta, P., Bhatia, G., Maurya, R., Nath, C., & Palit, G. 2005. Adaptogenic and anti-amnesic properties of *Evolvulus alsinoides* in rodents. *Pharmacology, Biochemistry, and Behavior* 81:424–432.

Slob, A. K., Koster, J., Radder, J. K., & van der Werff ten Bosch, J. J. 1990. Sexuality and psychophysiological functioning in women with diabetes mellitus. *Journal of Sex and Marital Therapy* 16:59–69.

Small, D. M., Zatorre, R. J., Dagher, A., Evans, A. C., & Jones-Gotman, M. 2001. Changes in brain activity related to eating chocolate: from pleasure to aversion. *Brain* 124:1720–1733.

Smith, D. E., Wasson, D. R., & Apter-Marsh, M. 1984. Cocaine- and alcohol-induced sexual dysfunction in patients with addictive disease. *Journal of Psychoactive Drugs* 16:359–361.

Smith, E. R., Damassa, D. A., & Davidson, J. M. 1977. Plasma testosterone and sexual behavior following intracerebral implantation of testosterone propionate in the castrated male rat. *Hormones and Behavior* 8:77–87.

Smith, P. E., & Engle, E. T. 1927. Experimental evidence regarding the role of the anterior pituitary in the development and regulation of the genital system. *American Journal of Anatomy* 40:159–217.

Soulairac, A., & Soulairac, M. L. 1957. Action de l'amphetamine, de noradrenaline et de l'atropine sur le comportement sexual du rat male. *Journal of Physiology, Paris* 49:381–385.

Sovner, R. 1983. Anorgasmia associated with imipramine but not desipramine: case report. *Journal of Clinical Psychiatry* 44:345–346.

Spector, M. P., & Carey, M. P. 1990. Incidence and prevalence of sexual dysfunctions: a critical review. *Archives of Sexual Behavior* 19:389–409.

Spinella, M. 2001. *The Psychopharmacology of Herbal Medicines*. Cambridge, MA: MIT Press.

Stahl, S. M. 1999. Conventional neuroleptic drugs for schizophrenia and novel antipsychotic agents. In *Essential Psychopharmacology.* Cambridge: Cambridge University Press.

Starkman, M. N., & Schteingart, D. E. 1981. Neuropsychiatric manifestations of patients with Cushing's syndrome: relationship to cortisol and adrenocorticotropic hormone levels. *Archives of Internal Medicine* 141:215–219.

Steers, W. D. 2000. Neural pathways and central sites involved in penile erection: neuroanatomy and clinical implications. *Neuroscience and Biobehavioral Reviews* 24:507–516

Steidle, C., Schwartz, S., Jocoby, K., Sebree, T., Smith, T., & Bachand, R. 2003. AA2500 testosterone gel normalizes androgen levels in aging males with improvements in body composition and sexual function. *Journal of Clinical Endocrinology and Metabolism* 88:2673–2681.

Stein, E. A., Pankiewicz, J., Harsch, H. H., Cho, J. K., Fuller, S. A., Hoffmann, R. G., Hawkins, M., Rao, S. M., Bandettini, P. A., & Bloom, A. 1998. Nicotine-induced limbic cortical activation in the human brain: a functional MRI study. *American Journal of Psychiatry* 155:1009–1015.

Steinach, E. 1940. *Sex and Life.* New York: Viking.

Steinach, E., & Peczenik, O. 1936. Diagnostischer Test fur hormone-bedingre Storungen der mannlichen sexualfunktion und seineklinische Anwendung. *Wiener Klinische Wochenschrift* 49:388.

Steinman, J. L., Komisaruk, B. R., Yaksh, T. L., & Tyce, G. M. 1983. Spinal cord monoamines modulate the antinociceptive effects of vaginal stimulation in rats. *Pain* 16:155–166.

Stenager, E., Stenager, E. N., Jensen, K., & Boldsen, J. 1990. Multiple sclerosis: sexual dysfunctions. *Journal of Sex Education and Therapy* 16:262–269.

Stoleru, S., Gregoire, M.-C., Gerard, D., Decety, J., Lafarge, E., Cinotti, L., Lavenne, F., Le Bars, D., Vernet-Maury, E., Rada, H., Collet, C., Mazoyer, B., Forest, M. G., Magnin, F., Spira, A., & Comar, D. 1999. Neuroanatomical correlates of visually evoked sexual arousal in human males. *Archives of Sexual Behavior* 28:1–21.

Stone, C. P. 1932. The retention of copulatory ability in male rabbits following castration. *Journal of Genetic Psychology* 40:296–305.

Stone, C. P. 1938. Loss and restoration of copulatory activity in adult male rats following castration and subsequent injections of testosterone propionate. *Endocrinology* 23:529.

Suckling, J., Lethaby, A., & Kennedy, R. 2003. Local oestrogen for vaginal atrophy in postmenopausal women. *Cochrane Database of Systematic Reviews* 4:CD001500.

Sugrue, D. P., & Whipple, B. 2001. The Consensus-Based Classification of Female Sexual Dysfunction: barriers to universal acceptance. *Journal of Sex and Marital Therapy* 27:221–226.

Swaab, D. F., & Fliers, E. 1985. A sexually dimorphic nucleus in the human brain. *Science* 228:1112–1115.

Swanson, L. W. 1999. Limbic system. In *Encyclopedia of Neuroscience,* ed. G. Adelman & B. H. Smith. New York: Elsevier.

Szechtman, H., Adler, N. T., & Komisaruk, B. R. 1985. Mating induces pupillary dilatation in female rats: role of pelvic nerve. *Physiology and Behavior* 35:295–301.

Tagliamonte, A., Fratta, W., & Gessa, G. L. 1974. Aphrodisiac effect of L-DOPA and apomorphine in male sexually sluggish rats. *Experientia* 30:381–382.

Talbot, J. D., Marrett, S., Evans, A. C., Meyer, E., Bushnell, M. C., & Duncan, G. H. 1991. Multiple representations of pain in human cerebral cortex. *Science* 251:1355–1358.

Terzian, H., & Dalle Ore, G. 1955. Syndrome of Kluver and Bucy: reproduced in man by bilateral removal of the temporal lobes. *Neurology* 5:373–380.

Tetel, M. J., Getzinger, M. J., & Blaustein, J. D. 1993. Fos expression in the rat brain following vaginal-cervical stimulation by mating and manual probing. *Journal of Neuroendocrinology* 5:397–404.

Thorek, M. 1924. Experimental investigation of the role of the Leydig, seminiferous, and sertolic cells and effects of testicular transplantation. *Endocrinology* 8:61–90.

Thornhill, R., Gangestad, S. W., & Comer, R. 1995. Human female orgasm and mate fluctuating asymmetry. *Animal Behavior* 50:1601–1615.

Tiihonen, J., Kuikka, J., Kupila, J., Partanen, K., Vainio, P., Airaksinen, J., Eronen, M., Hallikainen, T., Paanila, J., Kinnunen, I., & Huttunen, J. 1994. Increase in cerebral blood flow of right prefrontal cortex in man during orgasm. *Neuroscience Letters* 170:241–243.

Timmers, R. L., Sinclair, L. G., & James, R. A. 1976. Treating goal-directed intimacy. *Social Work* 401–402.

Travers, S. P., & Norgren, R. 1995. Organization of orosensory responses in the nucleus of the solitary tract of rat. *Journal of Neurophysiology* 73:2144–2162.

Turna, B., Apaydin, E., Semerci, B., Altai, B., Cikili, N., & Nazli, O. 2005. Women with low libido: correlation of decreased androgen levels with female sexual function index. *International Journal of Impotence Research* 17:148–153.

Udry, J. R., & Morris, N. M. 1968. Distribution of coitus in the menstrual cycle. *Nature, London* 227:593–596.

Uitti, R. J., Tanner, C. M., Rajput, S. H., Goetz, C. G., Klawans, H. L., & Thiessen, B. 1989. Hypersexuality with antiparkinsonian therapy. *Clinical Neuropharmacology* 12:375–383.

Utian, W. H. 1975. Effect of hysterectomy, oophorectomy and estrogen therapy on libido. *International Journal of Obstetrics and Gynecology* 84:4314–4315.

Vale, J. 1999. Ejaculatory dysfunction. *BJU International* 83:557–563.

Valenstein, E. S. 1973. *Brain Control: A Critical Examination of Brain Stimulation and Psychosurgery.* New York: John Wiley and Sons.

Vance, E. B., & Wagner, N. N. 1976. Written descriptions of orgasm: a study of sex differences. *Archives of Sexual Behavior* 5:87–98.

Van der Schoot, D. K. E., & Ypma, A. F. G. V. M. 2002. Seminal vesiculectomy to resolve defecation-induced orgasm. *BJU International* 90: 761–762.

Van Geelen, J. M., van de Weijer, P. H., & Arnolds, H. 1996. Urogenital symptoms and their resulting discomfort in non-institutionalized 50–75-year-old Dutch women. *Nederlands Tijdschrift voor Geneeskunde* 140:713–716.

Van Goozen, S. H. M., Wiegant, V. M., Endert, E., Helmond, F. A., & Van de Poll, N. E. 1997. Psychoendocrinological assessment of the menstrual cycle: the relationship between hormones, sexuality and mood. *Archives of Sexual Behavior* 26:359–382.

Veening, J. G., & Coolen, L. M. 1998. Neural activation following sexual behavior in the male and female rat brain. *Behavioural Brain Research* 92:181–193.

Veronneau-Longueville, F., Rampin, O., Freund-Mercier, M.-J., Tang, Y., Calas, A., Marson, L., McKenna, K. E., Stoeckel, M.-E., Benoit, G., & Giuliano, F. 1999. Oxytocinergic innervation of autonomic nuclei controlling penile erection in the rat. *Neuroscience* 93:1437–1447.

Vinik, A., & Richardson, D. 1998. Erectile dysfunction in diabetes. *Diabetes Reviews* 6:16–33.

Vogt, B. A. 1999. Cingulate cortex. In *Encyclopedia of Neuroscience,* ed. G. Adelman & B. H. Smith. New York: Elsevier.

Waldinger, M. D., Zwinderman, A. H., & Olivier, B. 2003. Antidepressants and ejaculation: a double blind, randomized, fixed-dose study with mirtazapine and paroxetine. *Journal of Clinical Psychopharmacology* 23:467–470.

Wallen, K., Winston, S., Gaventa, M., Davis-Dasilva, M., & Collins, D. C. 1984. Periovulatory changes in female sexual behavior and patterns of ovarian steroid secretion in group living rhesus monkeys. *Hormones and Behavior* 29:322–337.

Wang, C., Cunningham, G., Dobs, A., Iranmanesh, A., Matsumoto, A. M., Snyder, P. J., Weber, T., Berman, N., Hull, L., & Swerdloff, R. S. 2004. Long term testosterone gel (Andro-Gel) treatment maintains beneficial effects on sexual function and mood, lean and fat mass, and bone mineral density in hypogonadal men. *Journal of Clinical Endocrinology and Metabolism* 89:2085–2098.

Warner, M. D., Peabody, C. A., & Whiteford, H. A. 1987. Trazodone and priapism. *Journal of Clinical Psychiatry* 48:244–245.

Waters, C., & Smolowitz, J. 2005. Impaired sexual function. In *Parkinson's Disease and Nonmotor Dysfunction,* ed. R. F. Pfeiffer. New York: Humana Press.

Waxenburg, S. E., Drellich, M. G., & Sutherland, A. M. 1959. The role of hormones in human behavior. I: Changes in female sexuality after adrenalectomy. *Journal of Clinical Endocrinology* 19:193–202.

Waxenburg, S. E., Frinkheimer, J. A., Drellich, M. G., & Sutherland, A. M. 1960. The role of hormones on human behavior. II: Changes in sexual behavior in relation to vaginal smears of breast cancer patients with oophorectomy and adrenalectomy. *Psychosomatic Medicine* 22: 435–442.

Weder, B., Azari, N. P., Knorr, U., Seitz, R. J., Keel, A., Nienhusmeier, M., Maguire, R. P., Leenders, K. L., & Ludin, H.-P. 2000. Disturbed functional brain interactions underlying deficient tactile object discrimination in Parkinson's disease. *Human Brain Mapping* 11:131–145.

Weeks, D. J. 2002. Sex for the mature adult: health, self-esteem and countering ageist stereotypes. *Sexual and Relationship Therapy* 17:231–240.

Weijmar Schultz, W. C. M., van de Wiel, H. B. M., Klatter, J. A., Sturm, B. E., & Nauta, J. 1989. Vaginal sensitivity to electric stimuli: theoretical and practical implications. *Archives of Sexual Behavior* 18:87–95.

Wermuth, L., & Stenager, E. 1995. Sexual problems in young patients with Parkinson's disease. *Acta Neurologica Scandinavica* 91:453–455.

Wersinger, S. R., Baum, M. J., & Erskine, M. S. 1993. Mating-induced Fos-like immunoreactivity in the rat forebrain: a sex comparison and a dimorphic effect of pelvic nerve transection. *Journal of Neuroendocrinology* 5:557–568.

Whipple, B. 1987. Sexual counseling of couples after a mastectomy or a myocardial infarction. *Nursing Forum* 23:85–91.

Whipple, B. 1990. Female sexuality. In *Sexual Rehabilitation of the Spinal-Cord-Injured Patient,* ed. J. F. Leyson. Clifton, NJ: Humana Press.

Whipple, B. 2000. *Guide to Healthy Living for Men and Those Who Love Them.* New York: Pfizer. Videotape.

Whipple, B. 2002a. Does the partner have a role in the treatment of erectile dysfunction? Paper presented at 10th World Congress of the International Society for Sexual and Impotence Research, Montreal. September 22–26.

Whipple, B. 2002b. Women's sexual pleasure and satisfaction: a new view of female sexual function. *Female Patient* 27:39–44.

Whipple, B. 2005. *Lecture on Sexuality in Mid-life and Beyond.* Montreal: World President's Organization.

Whipple, B., & Brash-McGreer, K. 1997. Management of female sexual dysfunction. In *Sexual Function in People with Disability and Chronic Illness: A Health Professional's Guide,* ed. M. L. Sipski & C. J. Alexander. Gaithersburg, MD: Aspen.

Whipple, B., Gerdes, C. A., & Komisaruk, B. R. 1996. Sexual response to self-stimulation in women with complete spinal cord injury. *Journal of Sex Research* 33:231–240.

Whipple, B., & Gick, R. 1980. A holistic view of sexuality: education for the professional. *Topics in Clinical Nursing* 1:91–98.

Whipple, B., &. Komisaruk, B. R. 1985. Elevation of pain threshold by vaginal stimulation in women. *Pain* 21:357–367.

Whipple, B., & Komisaruk, B. R. 1988. Analgesia produced in women by genital self-stimulation. *Journal of Sex Research* 24:130–140.

Whipple, B., & Komisaruk, B. R. 1997. Sexuality and women with complete spinal cord injury. *Spinal Cord* 35:136–138.

Whipple, B., & Komisaruk, B. R. 2002. Brain (PET) responses to vaginal-cervical self-stimulation in women with complete spinal cord injury: preliminary findings. *Journal of Sex and Marital Therapy* 28:79–86.

Whipple, B., Myers, B., & Komisaruk, B. R. 1998. Male multiple ejaculatory orgasms: a case study. *Journal of Sex Education and Therapy* 23:157–162.

Whipple, B., Ogden, G., & Komisaruk, B. R. 1992. Physiological correlates of imagery induced orgasm in women. *Archives of Sexual Behavior* 21:121–133.

Whipple, B., & Scura, K. W. 1989. HIV and the older adult: taking the necessary precautions. *Journal of Gerontological Nursing* 15:15–19.

Wildt, L., Kissler, S., Licht, P., & Becker, W. 1998. Sperm transport in the human female genital tract and its modulation by oxytocin as assessed by hysterosalpingoscintigraphy, hysterotonography, electrohysterography and Doppler sonography. *Human Reproduction Update* 4:655–666.

Wilson, G. T. 1981. The effects of alcohol on human sexual behavior. In *Advances in Substance Abuse: Behavioral and Biological Research,* vol. 2, ed. N. K. Mells. Greenwich, CT: JAI Press.

Wincze, J. P., Albert, A., & Bansal, S. 1993. Sexual arousal in diabetic females: physiological and self-report measures. *Archives of Sexual Behavior* 22:587–601.

Wirshing, D. A., Pierre, J. M., Marder, S. R., Saunders, C. S., & Wirshing, W. C. 2002. Sexual side effects of novel antipsychotic medications. *Schizophrenia Research* 56:25–30.

Woodrum, S. T., & Brown, C. S. 1998. Management of SSRI-induced sexual dysfunction. *Annals of Pharmacotherapy* 32:1209–1215.

Woods, N. F. 1984. *Human Sexuality in Health and Illness.* St. Louis: CV Mosby.

World Health Organization. 1992. *International Statistical Classification of Disease and Related Health Problems*. Geneva: World Health Organization.

Yang, C. C., Bowen, J. R., Kraft, G. H., Uchio, E. M., & Kromm, B. G. 2000. Cortical evoked potentials of the dorsal nerve of the clitoris and female sexual dysfunction in multiple sclerosis. *Journal of Urology* 164:2010–2013.

Zemishlany, Z., Aizenberg, D., & Weizman, A. 2001. Subjective effects of MDMA ("Ecstasy") on human sexual function. *European Psychiatry* 16:127–130.

Zesiewicz, T. A., Heilal, M., & Hauser, R. A. 2001. Sildenafil citrate (Viagra) for the treatment of erectile dysfunction in men with Parkinson's disease. *Movement Disorders* 16:305–308.

Zifa, E., & Fillion, G. 1992. 5-HT receptors. *Pharmacological Reviews* 44:401–458.

Zivadinov, R., Zorzon, M., Locatelli, L., Stival, B., Monti, F., Nasuelli, D., Tommasi, M. A., Bratina, A., & Cazzato, G. 2003. Sexual function in multiple sclerosis: a MRI, neurophysiological and urodynamic study. *Journal of Neurological Science* 210:73–76.

Zumpe, D., Bonsall, R. W., & Michael, R. P. 1993. Effects of the nonsteroidal aromatase inhibitor, fadrazole, on the sexual behavior of male cynomolgus monkeys (*Macaca fascicularis*). *Hormones and Behavior* 27:200–215.

Index

About the Authors

Barry R. Komisaruk, Ph.D., is a professor in the Department of Psychology, Rutgers University, and an adjunct professor in the Department of Radiology at the University of Medicine and Dentistry of New Jersey. Supported by federal, state, and foundation grant funding, he has published over 145 research papers and 3 books.

Carlos Beyer-Flores, Ph.D., is head of the Laboratorio Tlaxcala–CINVESTAV and of CIRA–Universidad Autónoma de Tlaxcala, Mexico. He has published over 200 research papers and was awarded the coveted National Prize of the Mexican Academy of Sciences.

Beverly Whipple, Ph.D., R.N., FAAN, is a certified sex educator, sex counselor, and sex researcher and Professor Emerita at Rutgers University. She has published over 160 research papers and co-authored 6 books, one of which, *The G Spot,* has been translated into 19 languages and was an international bestseller.